THE FOCAL EASY G

CAKEWALK SONAR

The Focal Easy Guide Series

Focal Easy Guides are the best choice to get you started with new software, whatever your level. Refreshingly simple, they do not attempt to cover everything, focusing solely on the essentials needed to get immediate results.

Ideal if you need to learn a new software package quickly, the Focal Easy Guides offer an effective, time-saving introduction to the key tools, not hundreds of pages of confusing reference material. The emphasis is on quickly getting to grips with the software in a practical and accessible way to achieve professional results.

Highly, illustrated in color, explanations are shot and to the point. Written by professionals in a user-friendly style, the guides assume some computer knowledge and an understanding of the general concepts in the area covered, ensuring they aren't patronizing!

Series editor: Rick Young (www.digitalproduction.net)

Director and Founding Member of the UK Final Cut User Group, Apple Solutions Expert and freelance television director/editor, Rick has worked for the BBC, Sky, ITN, CNBC and Reuters. Also a Final Cut Pro Consultant and author of the best-selling *The Easy Guide to Final Cut Pro*.

Titles in the series:

The Easy Guide to Final Cut Pro 3, **Rick Young**

The Focal Easy Guide to Final Cut Pro 4, **Rick Young**

The Focal Easy Guide to Final Cut Express, **Rick Young**

The Focal Easy Guide to Maya 5, **Jason Patnode**

The Focal Easy Guide to Discreet Combustion 3, **Gary M. Davis**

The Focal Easy Guide to Premiere Pro, **Tim Kolb**

The Focal Easy Guide to Flash MX 2004, **Birgitta Hosea**

THE FOCAL EASY GUIDE TO CAKEWALK SONAR

For new users and professionals

TREV WILKINS

AMSTERDAM • BOSTON • HEIDELBERG • LONDON • NEW YORK • OXFORD
PARIS • SAN DIEGO • SAN FRANCISCO • SINGAPORE • SYDNEY • TOKYO
Focal Press is an imprint of Elsevier

Focal Press
An imprint of Elsevier
Linacre House, Jordan Hill, Oxford OX2 8DP
30 Corporate Drive, Burlington, MA 01803

First published 2005

British Library Cataloguing in Publication Data
A catalogue record for this book is available from the British Library

Library of Congress Cataloguing in Publication Data
A catalogue record for this book is available from the Library of Congress

ISBN 0 240 51975 2

For information on all Focal Press publications visit our website at:
www.focalpress.com

Typeset by Charon Tec Pvt. Ltd, Chennai, India
Printed and bound in Italy

Contents

Preface

Cakewalk's flagship software is SONAR designed for the PC.

It is very powerful for recording and editing both audio and MIDI.

It is intuitive and very easy to use, particularly if you are familiar with Windows based applications.

It is easily capable of very high quality results that are as good as any other platform currently in use, the only proviso here is the quality of the hardware that you use with it.

Anyone who is looking to move up from a tape based studio will find many similarities but you can throw away those razor blades as the SONAR 'scalpel' is controlled on screen and if you make a mistake it can be easily undone – hi-tech surgery indeed.

The number of tracks used to be a prime consideration when recording as there was a finite limit imposed by tape and mixing was limited to how many channels the mixer and recorder had but there are no such limits with SONAR. The track count is unlimited (well at least until your PC can't take any more) and the mixing console will handle as many channels as the computer will.

Some people expect computer music to consist of bleeps and clicks but this is not the case at all as you can record delicate instruments such as a violin to the highest fidelity as well as overdriven guitars, percussion and vocals or even the ambience of the sea crashing on the rocks if you choose a portable laptop style computer. There are no limits.

SONAR is designed to enable recording and editing of multiple tracks, either MIDI, audio or any combination of them and is also fully compatible with most industry standard formats. It also supports the major soft synth and plug-in formats and contains it's own suite of synths and effects all ready to go at a price that is less than a mid-range mixing console.

Whatever your objectives SONAR will make your work easy and enjoyable with enough tools and features to cope with the most demanding tasks.

The aim of this book is to provide you with a good grounding in SONAR and its uses whether you are a complete novice or if you already have a background in music and recording. There is obviously a vast amount of knowledge that can't be covered in depth here so I have tried to keep within sensible boundaries but SONAR complies with normally used industry standards so if you decide to find out more about a certain area, MIDI for instance, then you can be sure that SONAR will match up with the generally accepted MIDI standard but also extends to cover other proprietary MIDI formats as well.

SONAR is a great package but as with any creative tool it is only as good as the material it has to work with, it is possible to improve recorded material but you must always bear in mind that a good performance will always sound better than a bad one, even after it has been tweaked, and it will also require much less tweaking!

There is obviously a learning curve but the aim of the book is to make it a gentle one; it is also made easier by the fact that SONAR uses many commands and keyboard shortcuts that Windows users will already be familiar with.

The layout of the book is designed so that you can read right through it from start to finish but also skip to any chapter or section relevant to your work at any time. I'd suggest taking time to read right through from the start first as it will give you a good grounding in the basics.

I've used the flagship Producer version of SONAR as the basis for this book but most of the techniques will also apply to the Studio version as well and a chapter at the end of the book outlines the differences between the two versions.

To make menu selections easier to follow I've taken up the often used method of listing the menu tree with a > symbol between each menu or submenu step like this . . .

File>Import>Audio

This means that you select the File menu then go to the Import selection and from the options here you choose Audio.

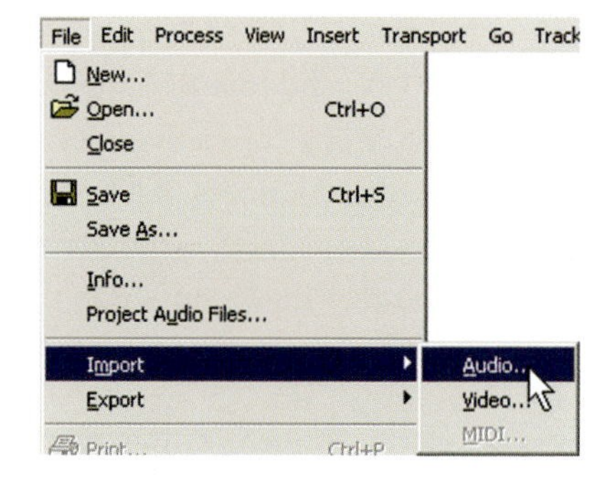

When a word is particularly relevant to SONAR I've used capital letters at the start of the word to show the importance of the word such as 'Track'. This also helps to show the context of the word as a Track in SONAR can be differentiated from an album track.

I've tried to keep to everyday language wherever possible but sometimes technical terms are unavoidable, in these cases I've tried to offer an alternative word as well to help you understand the meaning.

It's worth noting that SONAR has a very good Help system and the user forum at www.cakewalk.com/forum is also a mine of information should you have any problems. Cakewalk's technical support is also highly respected but before you contact them always make sure that you have the latest updates for SONAR (available from the Support pages) and your hardware drivers (particularly the soundcard drivers) are the most recent ones available.

Dedication

Much love and thanks to Shirley for putting up with me working in this business and especially while I've been writing this book as well.

A big thank you to Damien, Daniel, Christian, Jamie and all of my family for being there and understanding when I'm away.

Acknowledgements

Thanks to all at Cakewalk, particularly Morten Saether who checked the technical details and made some invaluable suggestions.

Everyone at Focal Press, especially Beth, Georgia, Marie, Stephanie and Emma for their patience and for helping me through this.

Tobias Thon at Native Instruments.

Steve and Craig at Minnetonka.

Antoni Ozynski at PSP Audioware.

All the Cakewalk users out there who always seem so enthusiastic and willing to share their knowledge with other users.

CHAPTER 1
SONAR BASICS

Some terms used throughout this book are specific to SONAR and so a little explanation will make life much easier for you as you work your way through it. I'll keep it simple and expand on things as the book progresses but for now here are a few things you should know . . .

The Main Track View

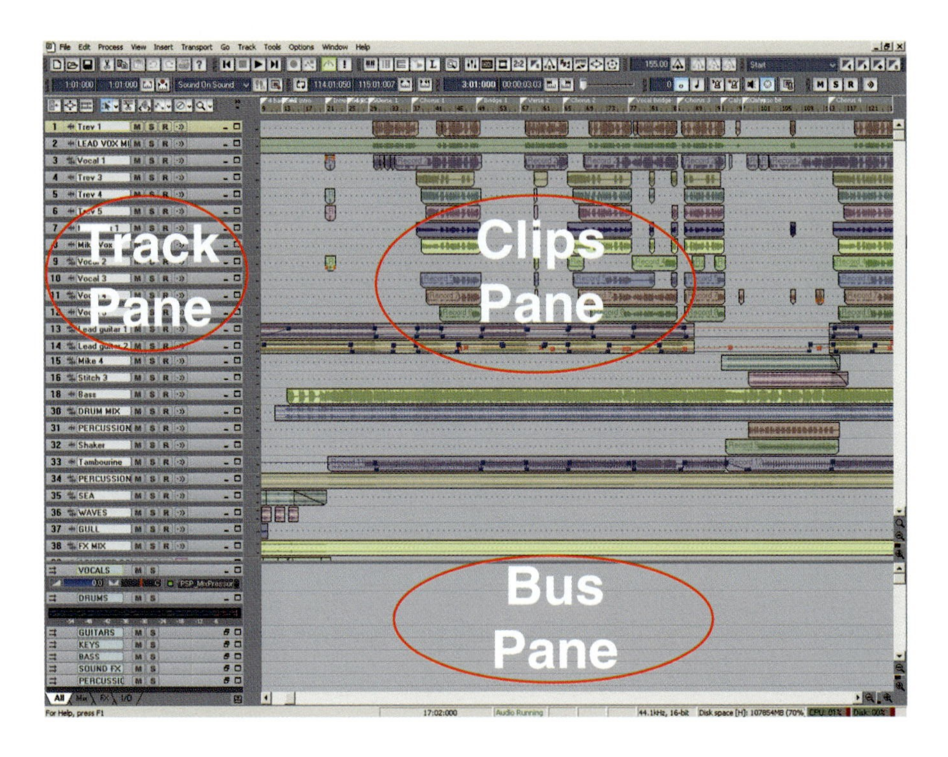

This is the View that you'll use most and it's split into several areas as you can see. Each main Pane can be resized by dragging the splitter bars when your cursor is held over it and looks like this ( ). Here's a brief description of each one:

1 The Track Pane: This is where the controls for each Track are situated. The type of controls can be selected by clicking one of the tabs at the bottom of the View (All / Mix / FX / I/O) in order to see All controls, Mix controls, FX (effects) controls and I/O (Input and Output) controls. I tend to use the All settings most.

2 Clips Pane: This is where the Clips, both audio and MIDI can be seen and edited. The song cursor (or Song Position Pointer) moves through this Pane as the song plays (this is referred to as the

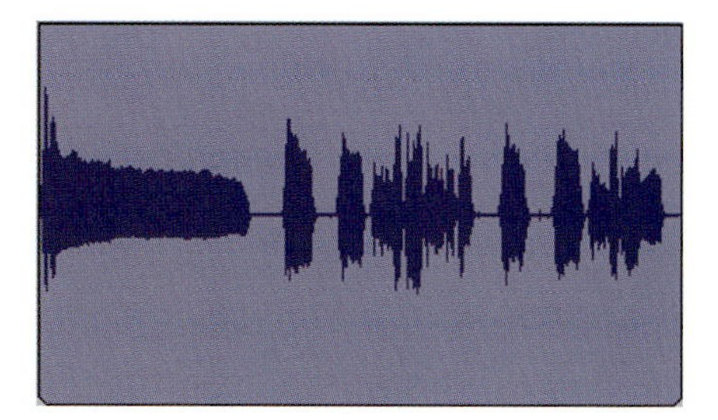
Audio clip

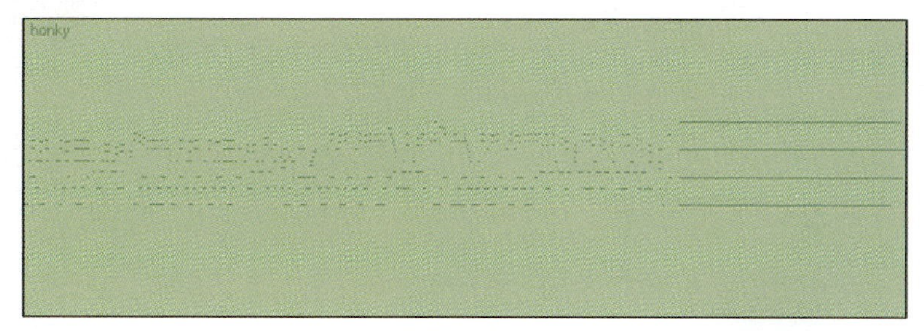

MIDI clip

current Now Time) and appears as a vertical line with this flag at the top ( ).

3 The Bus Pane: Can be shown or hidden by using this button ( ) and contains all current Buses as defined by you (Aux Buses, Subgroups, etc.).

4 Inspector: This is a channel strip for the currently selected Track, Bus or Main Output either audio or MIDI (the one that's the most recently highlighted) and it can be configured to show various features by use of the buttons at the bottom. It looks like this for audio and this for MIDI.

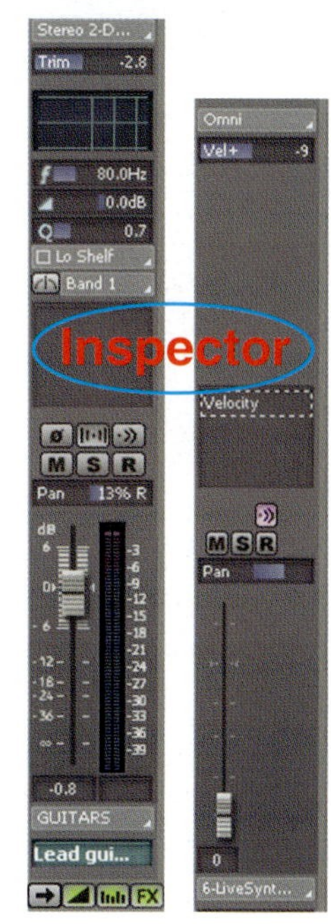

Audio MIDI

It is opened and closed by using this button from the tools at the top of the Track Pane ( ).

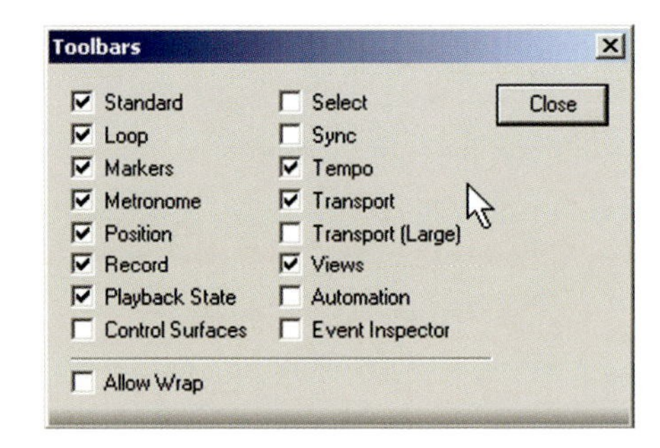

5 Toolbars: These can be shown or hidden, rearranged and dragged to float anywhere on the screen or docked at the top or bottom of your screen. Toolbars without any edit boxes can also be docked at the left or right of your screen. To make selections use the View>Toolbars menu to open the Toolbar Options dialog.

6 Navigator: An overview of the current Project's contents with a resizable box (the Track Rectangle) that can be used to update what's visible in the Clips Pane.

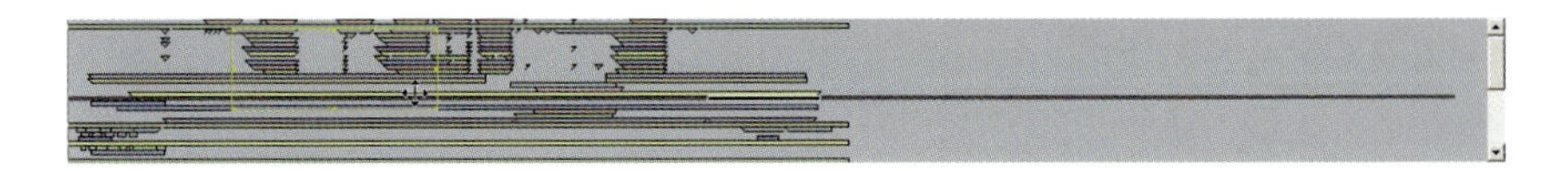

It is opened and closed using this button from the tools at the top of the Track Pane ( ).

7 Video Thumbnails Track: Shows thumbnails of any currently loaded video file and is opened and closed using this button from the tools at the top of the Track Pane ( ).

8 Tool selector: Tools most commonly used for editing and showing/hiding parts of the Track View. Hold your mouse over a button to see what it does and its associated shortcut key ( ).

9 Time Ruler: Shows the current position in time. The format can be changed by right-clicking on it and selecting another option from the Time Ruler Format choices. Any inserted Markers are also shown here.

10 Controls in the Track Pane.

These show the type of Track. Audio . . .

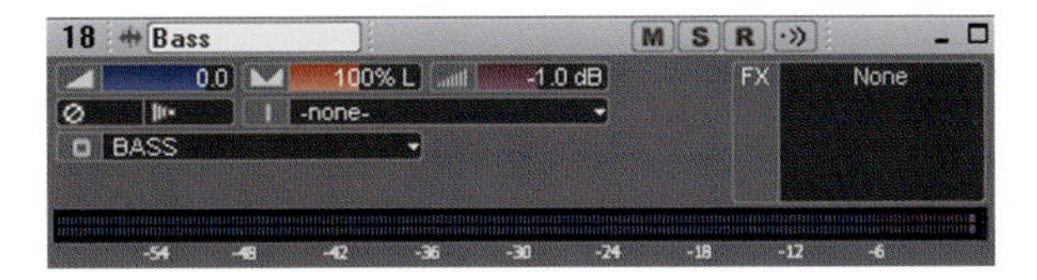

or MIDI . . .

and enable you to Mute (M) Solo (S) and Record Arm (R) the Track.

(-5.2) This type of readout after playing an audio Track is the maximum signal level reached. It can be reset by double-clicking on it.

As you progress you'll use other Views and tools associated with them but to start with it's best to keep things simple. If you've previously recorded using tape for instance then get used to how this translates into digital recording first so that you feel familiar with the controls before diving into soft synths or other alien topics.

Many options are obtained by right-clicking on specific areas, such as a Clip, and in general SONAR should be quite intuitive to anyone who is familiar with the Windows environment but if you're entirely new to all of this then don't be put off by the learning curve just make sure you can handle your computer before adding digital sound technology into the pot!

The extensive use of images in this book should help you on your way so if you don't quite understand the terminology just take a look at the pictures as it will probably become clearer when you do.

Undoing

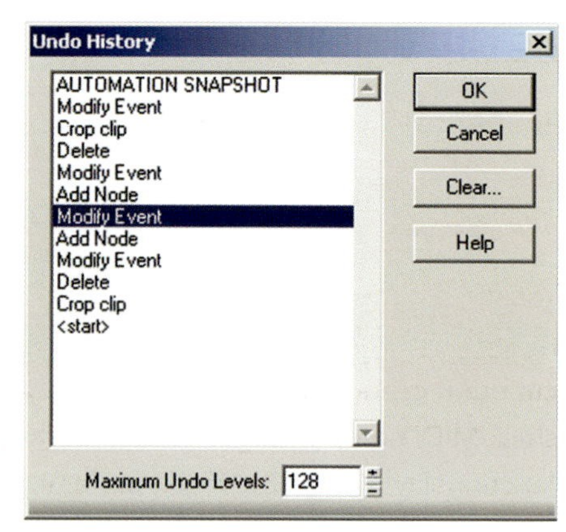

SONAR has the ability for you to undo many operations providing you haven't closed the Project down. You can use the common Ctrl+Z keyboard shortcut, click the Undo icon or choose Undo from the Edit menu. To go back several levels you should go to the Edit menu and select History, then you can choose which point you want to go back to, highlight it and click OK. You can even set the maximum number of levels that you can Undo!

Saving

The Save options are the same as most Windows-based applications but SONAR will allow you to save in several specific formats. Note that the Normal type of file doesn't contain your audio if you back it up. You should use Cakewalk Bundle files to make proper backup copies of your songs as these will contain all necessary files and data.

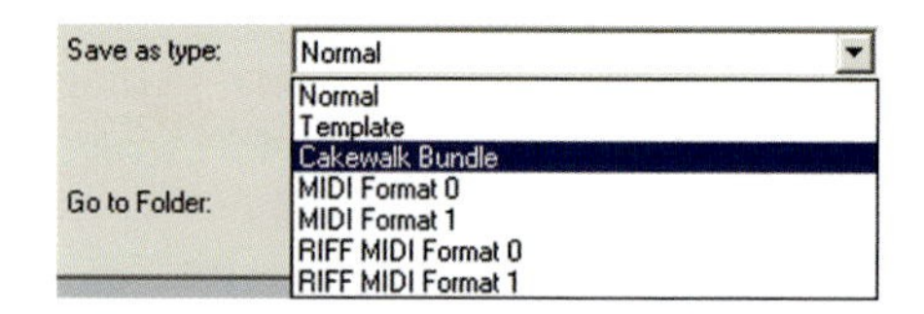

Tool Tips

A very helpful facility are the Tool Tips which appear when you hold the mouse cursor over a button or control. This will generally provide information about the item and also display the default keyboard shortcut.

Toolbars

A wide selection of Toolbars are available to cover many aspects of SONAR but I suggest that you don't clutter your screen up with too many of them initially as they can be confusing. You will need some however and I'd suggest the following.

Standard

Provides all of your essential Windows style tools ( ).

Transport

You'll need this to control the Playback, Recording, Rewind, etc. ( ). It also contains an audio engine trip ( ) which may cut out if overloaded and a panic button ( ) which can be clicked to release stuck MIDI notes. A larger version of the Transport Toolbar with more functionality is available if you prefer it.

Views

Fast access to alternative Views such as the Piano Roll or mixing Console ( ).

Tempo

Access to quickly change Tempo ( ).

Markers

When Markers ( ) are inserted you can change their names and times, jump to any one and also set a Project's Pitch here.

Loop

Enables you to play back a defined area continuously, useful for editing ( ). This button ( ) sets the loop to the current time selection.

Position

You can quickly jump to the beginning or end of the currently defined area or drag the slider to move around your song ( ).

Metronome

Set various parameters for metronome use including count-ins ( ).

Common terms and what they mean – They may be different colors but in general:

A mono audio Clip looks like this.

A stereo audio Clip looks like this.

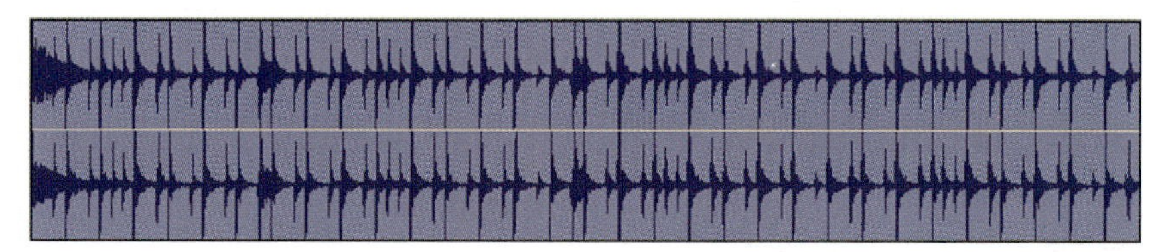

A MIDI Clip looks like this.

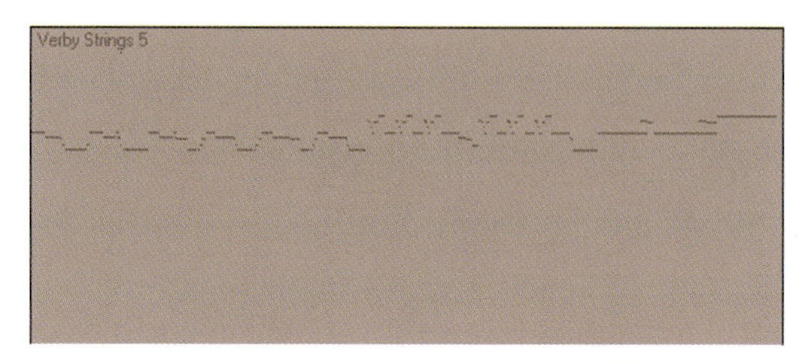

Values may be changed by typing in a new value ( ) or using the spinners ( ).

To change the name of a Track or Bus in the Track View double-click it and type in your new name ( ).

To change the name of a Track or Bus in the Console View double-click it and type in your new name.

A soft synth is a software version of a synthesizer such as the Native Instruments version of the DX7 hardware synth called FM7.

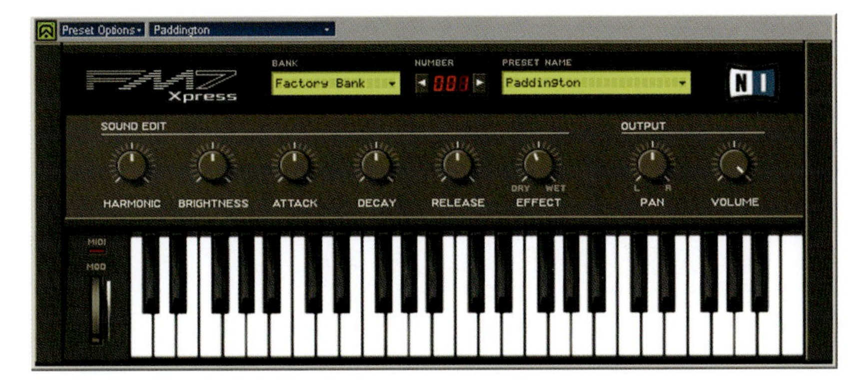

It could also be a software device that is designed for a specific job such as a sampler or drum synthesizer that only exists in software and has no real hardware counterpart at all.

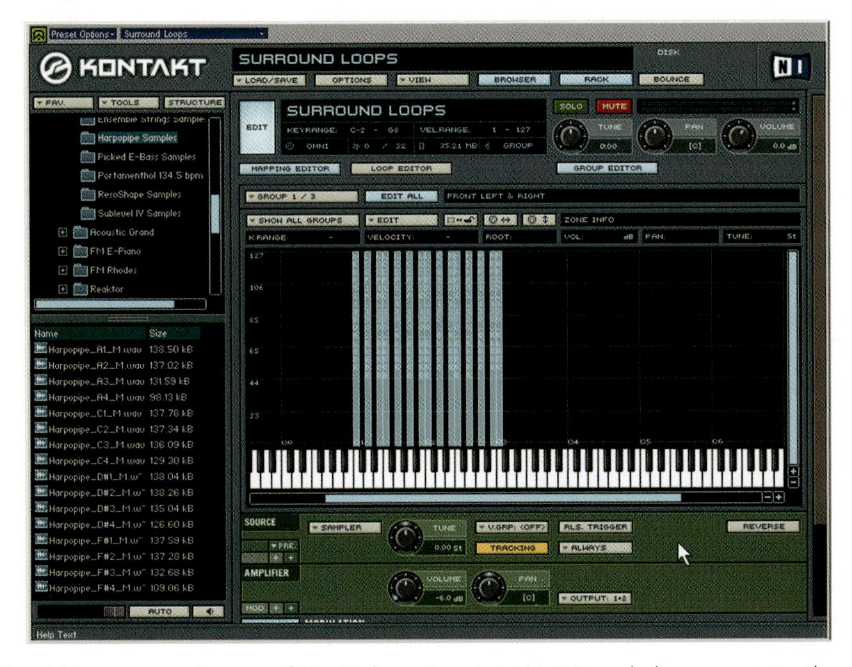

Don't forget to make use of the Help system in SONAR and always register the software so that you are eligible for technical support should you need it.

CHAPTER 2
SETTING UP

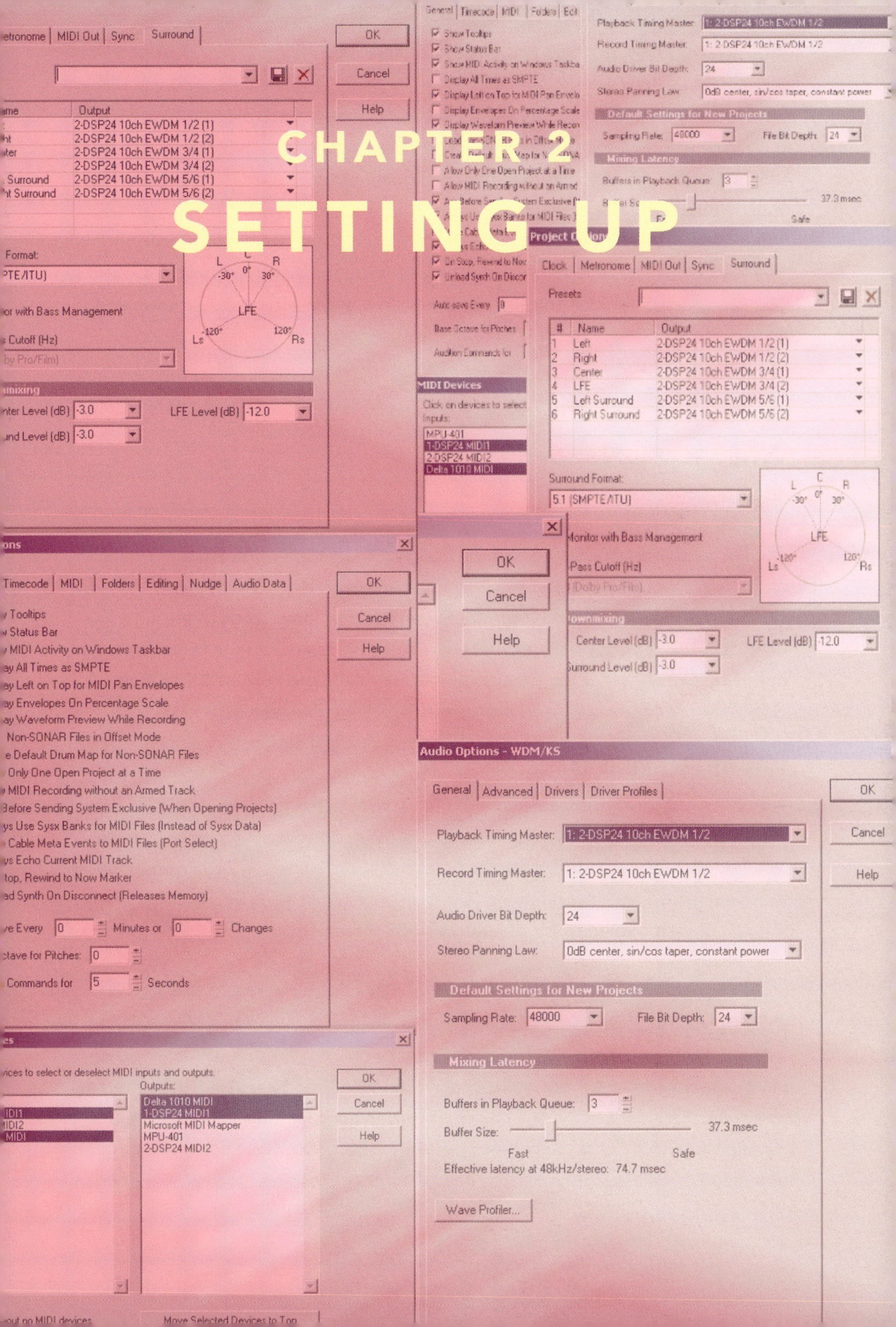

Setting up your system to work with SONAR is a fairly easy process, certainly as easy as working with any other similar sequencing package but there are a few things that will make a difference to the quality of your work and the speed at which you are able to carry it out.

Making good choices right at the start will make life smoother straight away but you can change virtually any setting at any time in the future should the need arise either because you want to tune the system a little or simply due to an upgrade of hardware.

What You'll Need

The Computer

The first and most important item on your list is obviously a suitable PC and this is relatively easy to define as Cakewalk provide the minimum system specifications on their product boxes and on their website at www.cakewalk.com. Usually any recent system will fall within the boundaries but just make sure that the processor, memory and hard drive(s) are at least above the minimum spec and if possible, higher. If you can you should try and meet or exceed Cakewalk's *recommended* specifications.

An extra hard drive dedicated to audio only will stream faster than a drive that shares the system or other programs and provide a higher number of possible simultaneous audio tracks. If you can, get a fast drive such as 7200 rpm or above for this purpose and if you have one with a large capacity this will obviate the need for deleting audio files too often.

Audio (G:)

If you're producing music then a CD writer (or burner) is a necessity both for creating your finished audio CDs and also for backing up your projects. A DVD writer is a wise investment as the extra capacity of DVD discs means that you can backup very large files that simply won't fit onto a CD and as most DVD writers will also write CDs they are the first choice if you can afford it. SONAR can generate Bundle files which contain all of the necessary data to recover a song including all of the audio but they do need a lot of space if they contain a lot of audio, often more than a CD can hold so a recordable DVD disc or even a removable hard drive will make life easier.

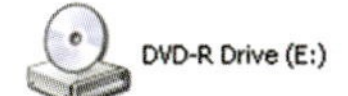

Apart from the usual ergonomic requirements like a good screen display, keyboard and mouse the most essential piece of kit for music making is a decent quality soundcard.

Soundcard Basics

The soundcard is the interface between your musical performance and the computer both on the way in and on the way out. If you record a vocal then it will go in via an input on the soundcard and come out via an output. If you play or record your MIDI keyboard it will send its data via the soundcard's MIDI input and can also send it to the soundcard's MIDI output as well as playing the music generated inside the computer (for instance by a soft synth) via an audio output.

Many computers have what are called 'onboard' soundcards which are basically a chip on the motherboard and a plate with the relevant sockets on, although they will do the job to some extent it is far better to use a dedicated soundcard suited to your specific needs and there are plenty to choose from. If possible then a soundcard capable of recording at 24 bit/96 kHz at least is recommended by Cakewalk, if you're unsure then check with your supplier.

Before buying a soundcard you need to consider exactly what you want to achieve both now and in the foreseeable future. The following choices will help you decide:

Will you be using a desktop or laptop PC?

Will you use an internal card (desktop) or an external device (desktop or laptop)?
If external then you need to know how it will connect to the PC, usually USB or Firewire.

Do you need MIDI input?
Maybe for a keyboard.

Do you need MIDI output?

To drive other MIDI equipment such as sound modules.

How many audio inputs and outputs do you need?

If you need to record more than two audio sources at the same time you'll need more inputs: if you want to mix externally or use external hardware effects you'll need more outputs. Usually the number of inputs and outputs matches so a card with eight inputs will have eight outputs.

Today's soundcards also have many high-end options such as built-in pre-amps, external converters, digital connectors and the like which can make life much easier or indeed more complicated! Careful thought before making a purchase will ensure that you get what you need and avoid adding complications by steering clear of unnecessary features.

A final important consideration is whether the soundcard has the correct drivers (the piece of software that enables the computer to 'talk' to the soundcard) and note that WDM or ASIO drivers are recommended by Cakewalk. Most popular cards have good quality drivers suitable for most applications but it's always worth checking compatibility with the manufacturer or Cakewalk by their websites or user forums. A good source of information for buyers is the music press as many regular publications run reviews and buyers guides covering all types of soundcards and related equipment, alternatively run a search on the Internet for buyers guides or music related websites as there are a vast amount of resources freely available.

Installation

SONAR comes with its own install routine built right in so all you need to do is insert the CD and follow the instructions. If the CD doesn't start then go to My Computer and double-click the CD drive's icon or right-click it and select Open. The installer will ask you a few questions as it proceeds and if you're not sure of an answer then it's usually alright to accept the default options, you can always change them later if you want to. You can run the installer at any time later and choose specific items to install even if you choose not to install them initially.

Choosing MIDI Ports

When first using SONAR you will be asked to choose MIDI Ports on your system (if it has any), which is a simple matter of selecting any input and output ports that you want to use. They will appear in SONAR as they appear here so if you want to reorder them then highlight one and use the 'Move Selected Devices to Top' button to move them around. This can be changed later. (See MIDI Options.)

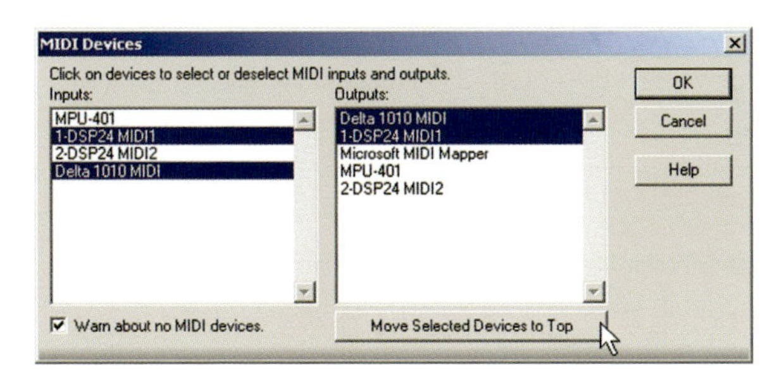

The Wave Profiler

The Wave Profiler will ask to run when you first fire up SONAR which looks a bit daunting but all it's doing is taking the hard work out of setting up your soundcard. It will check what the soundcard and its drivers are capable of and set SONAR up to work with it at an efficient level. If you change the driver, add another soundcard or change your existing one then you must run the Wave Profiler again. (See Audio Options.)

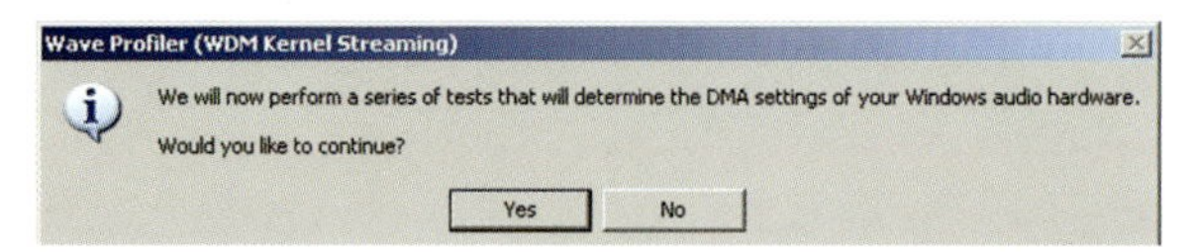

Customizing Options

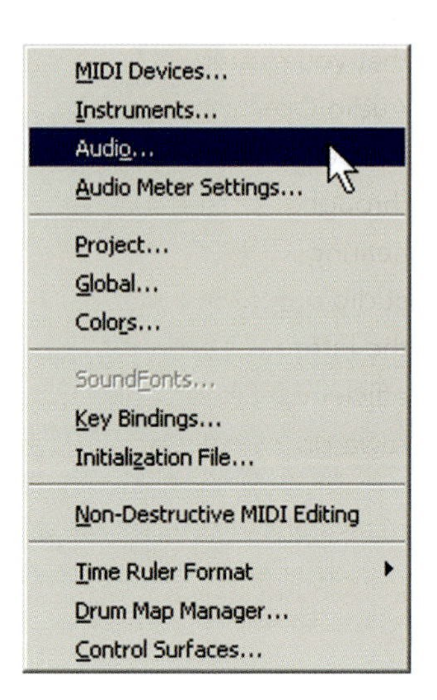

Once installed SONAR will usually work happily without any further adjustment but you may find that it's possible to tweak some of its settings to increase performance or for simple ergonomic reasons such as moving your audio storage to a new drive. Most of these adjustments can be carried out using the submenus from the main Options heading in SONAR and we'll come across some settings in their relevant chapters as we go through the book.

Options>MIDI Devices

This is where we can add, remove or change the order of available MIDI Ports on the system. Highlighted ones are active and can be moved to the top to change the order that they appear in by clicking the Move Selected Devices to Top button. The change is reflected wherever MIDI Ports appear throughout the program.

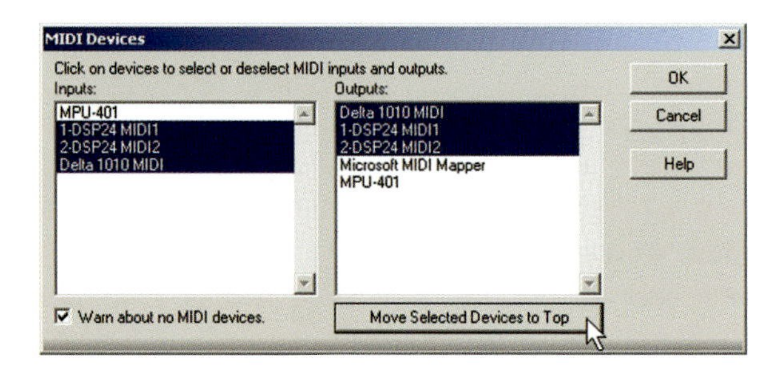

Options>Audio

This is where the important changes such as driver choice and latency can be made under the various tabs (General, Advanced, Drivers and Driver Profiles). If everything is running fine then it's wise to leave it that way but if not then you can change things around here but make a note of what you do so that you can go back to the earlier settings if your changes don't work out. The one setting that you may need to adjust on occasion is the Latency slider, although it's in the Audio Options dialog it will affect the latency experienced on your system which will be most noticeable when playing a MIDI instrument, particularly MIDI drums, through a soft synth when a lag or delay is felt between playing the note and hearing its sound; this is due to the time it takes for the soft synth to generate audio output in response to MIDI input. The latency slider can be dragged to the left to cut down on the delay but this will depend on the soundcard driver's efficiency. When you don't need low latency it's wise to have the slider more towards the right side and it's 'Safe' setting to prevent glitches and stutters.

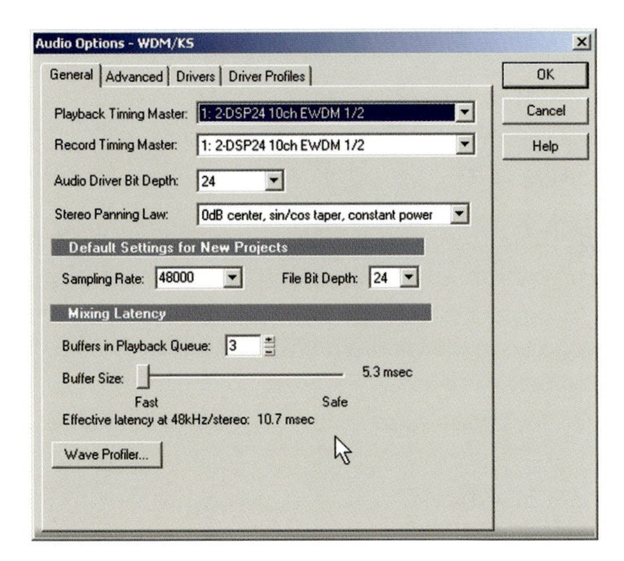

The Wave Profiler can be found here should you need it to reprofile the system at any time.

Many other options can be set here including the Sampling Rate and Bit Depth which we'll cover more in the Recording Audio section. (See Bits and Hertz.)

Options>Project

Your projects will probably have many settings in common but sometimes you might want to change certain aspects without affecting your usual style, this is where the Project Options dialog comes in. It has five tabs that cover Clock, Metronome, MIDI Out, Sync and Surround options that may be adjusted to suit, the settings will be saved with the project and automatically used the next time it is opened.

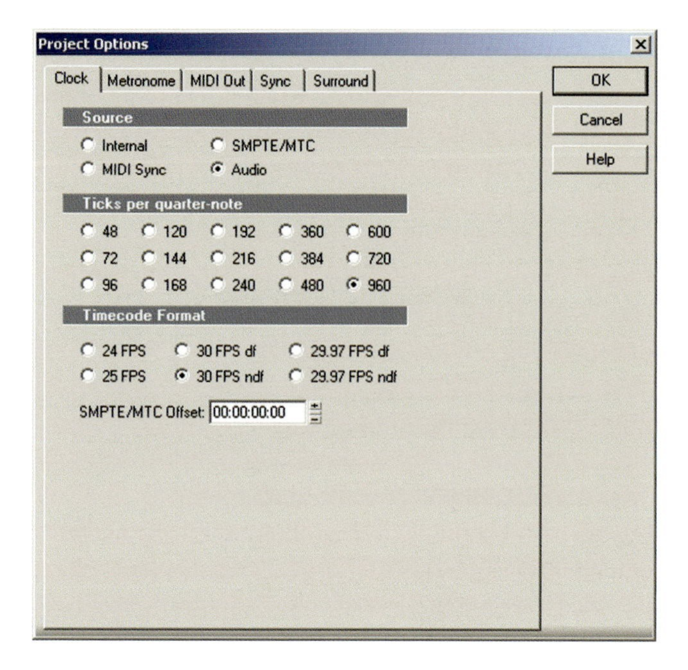

Options>Global

As the name implies these settings apply globally and will affect any projects used within SONAR. The default settings are sensible for most users and many of the choices on the General tab will be understandable to experienced users but again they are best left alone if you're not sure, you can always come back to them. One option you might like to use is the Auto-save facility as saving on a regular basis is vital to ensure that you don't lose that performance of a lifetime. Define the number of changes before a save or set a time limit if you don't trust yourself to do it manually.

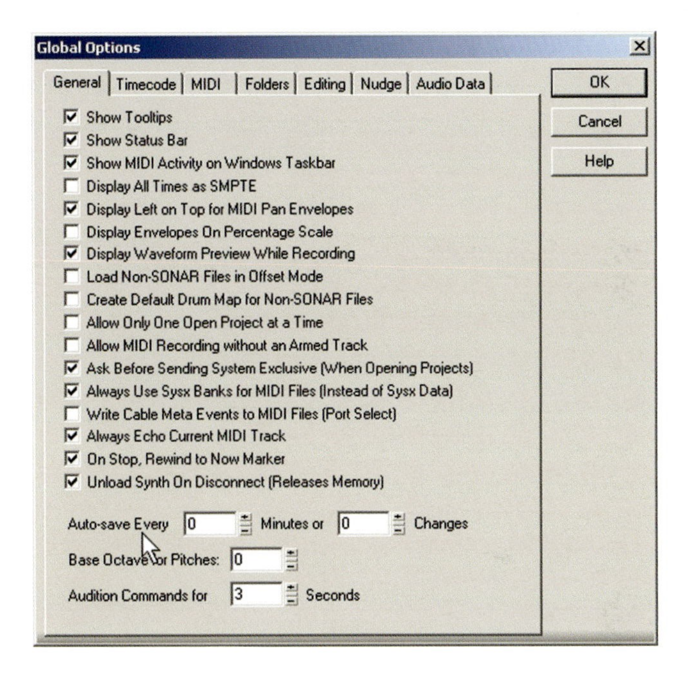

Timecode settings will only apply to users with external hardware that has to be synced up but the MIDI tab may apply to owners of instruments that generate such messages, you'll know if it applies to you or not. The Editing options will become more apparent as you progress and find your own way of working. Folders and Audio Data deserve a little more explanation . . .

Options>Global>Folders – These options define the default paths used when searching for various types of files stored on your system. You can specify the path here for each type which is particularly useful when streamlining your system or upgrading and adding new drives if your folders are moved.

The Wave Files field specifies the default path used when importing audio. Similarly if you use video then you may have a dedicated video drive and the default path for importing it can be specified in the same way using the Video files field.

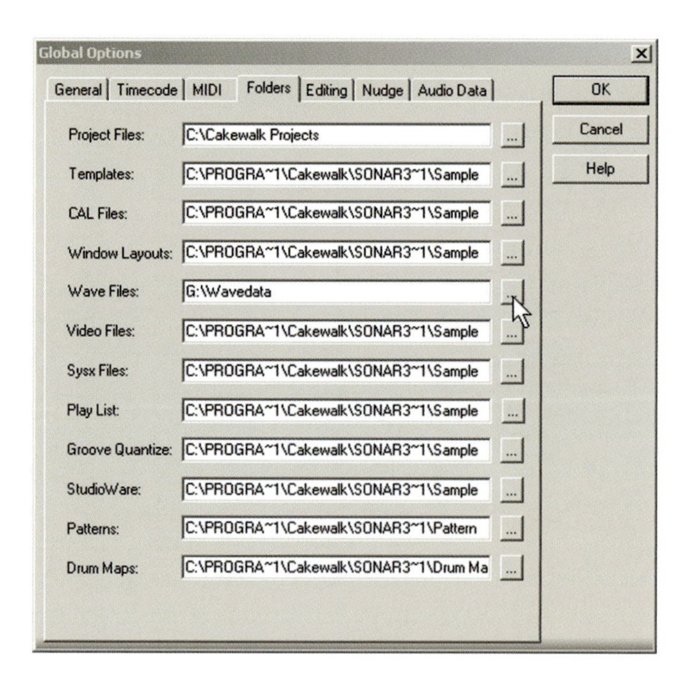

The Nudge tab provides options for setting three different Nudge values to Nudge Clips or MIDI notes by Musical Time values, Absolute Time values or to follow the current Snap To setting.

The Audio Data tab also has fields for the Global Audio Folder (your audio drive) and the Picture Folder which is where the actual pictures of your recorded audio waves are stored. The reason for the Picture Folder is that the next time you load up the Project the waves don't have to be drawn from scratch as their images are stored meaning faster loading times. If you keep this folder on a separate drive to the audio drive it won't interfere with your audio streaming.

Per-Project Audio Folders

This option is available from the Audio Data tab.

If you don't want to use the same drive every time for audio then you can tick this box and whenever you start a new project you will be asked where to store the data. SONAR will then name and direct all of your audio tracks to your specified folder which can ease backup and also be used if the same files are required by several users from a central location such as a server.

Optimization Tips

As I've already mentioned a separate hard drive for audio will really help the throughput and relieve the strain on your system drive but any upgrade will help.

Extra processing power will enable you to run more plug-in effects and synths.

More memory will enable you to load larger sample banks and also save the system using your hard drive as virtual memory.

Your graphics card might not seem important but faster graphics mean a smoother workflow and one with its own memory won't use up your system memory.

Try to keep your music PC as a dedicated machine without too many other applications onboard; if you can, use it solely for music and have another machine for games, Internet, etc. Alternatively you could use a dual-boot system but it is beyond the scope of this book to explain how to do this, try a Google search!

If you have any other tasks running such as virus scanners then turn them off.

Keep your system up to date by regularly checking for updates and new drivers on the Internet.

Turn off system sounds from 'Sounds and Audio Devices' in Control Panel. Under the Sounds tab choose 'No Sounds' as the scheme.

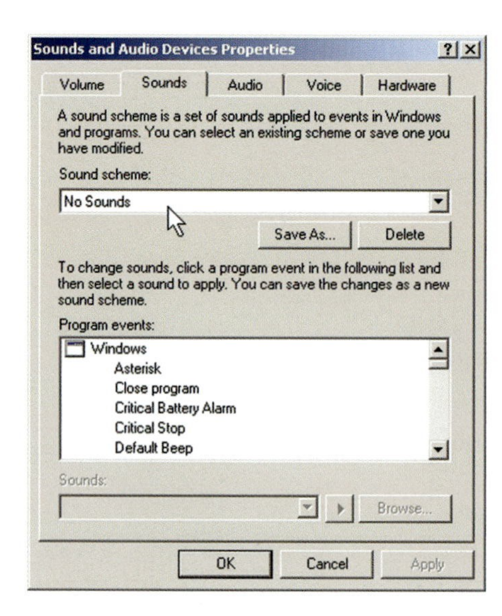

Right-click on the desktop, select Properties then under the Screen saver tab select (None) from the drop-down menu.

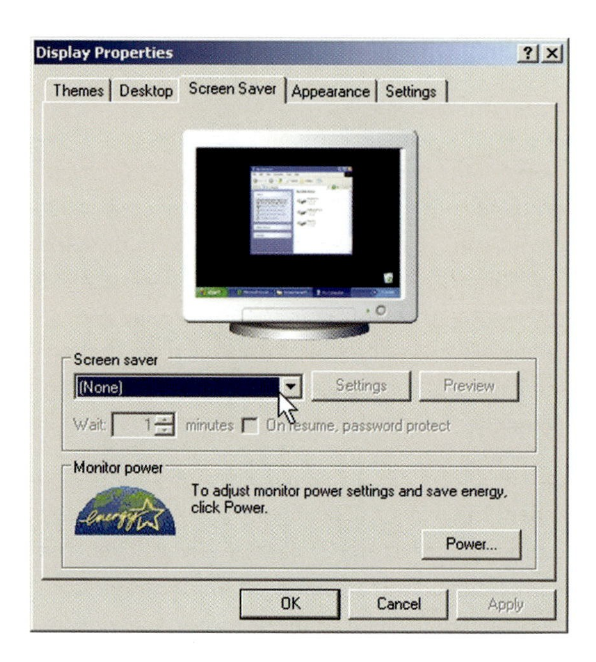

Turn off any unnecessary 'extras' such as visual effects like 'Use transition effects' or 'Show window contents while dragging'. Most of these items can be turned off from Windows Control Panel but if you are more experienced then you could try using the TweakUI utility from Microsoft.

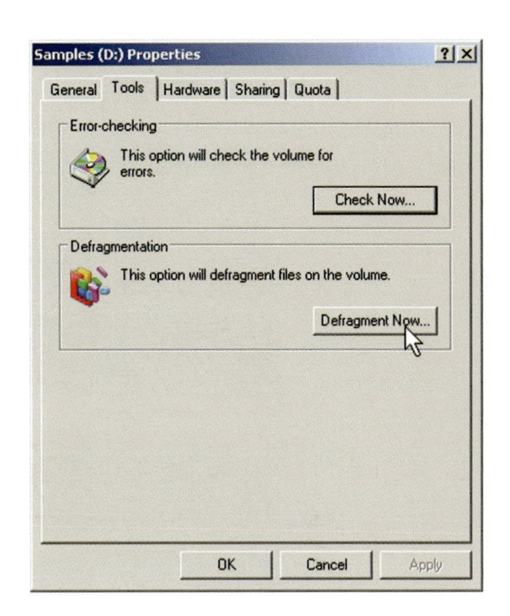

Make sure your hard drives are defragmented regularly, you can do this by opening My Computer then right-click on a drive's icon, choose Properties>Tools then click the Defragment Now button.

Some More Advanced Tweaks

In SONAR's Options>Audio dialog click the Advanced tab and under the Playback and Recording heading check out what driver types are available. You can try out WDM or ASIO to see which works best for you by selecting one then hitting OK, you'll then be reminded to restart SONAR for the changes to take effect. When you restart it you'll be prompted to run the Wave Profiler before it starts. If you have any glitches or problems using one driver then just switch to another one to see if matters improve. You'll probably find one type will be more efficient than the other.

On the General tab of the Audio Options you can select which device to use as the Playback and Record timing master. If you only have one device then there will be no choice but if you have more than one always select the best device; for instance a dedicated soundcard being preferable to an onboard sound chip. To deselect devices that you don't want to use, click on the Driver tab and click the device names to deselect them so that they aren't highlighted in blue. They then won't appear as available Inputs or Outputs in the program.

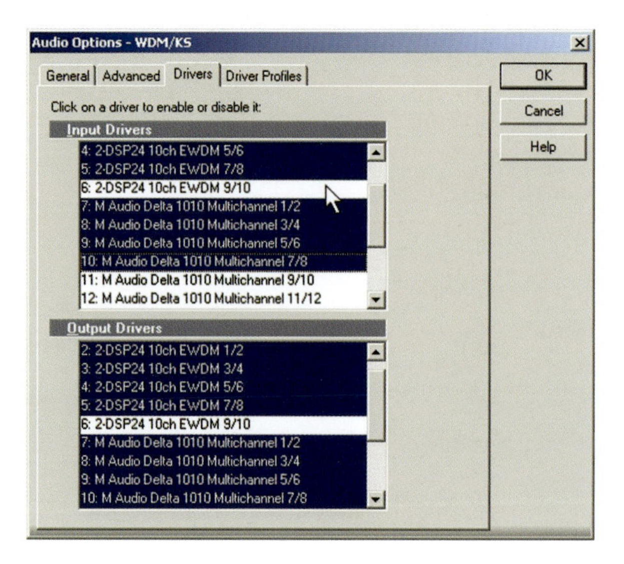

Templates

They're not very glamorous but they can be very useful and a real time-saver if you use a similar setup regularly. The basic method of creating a Template is to arrange all of the necessary features within SONAR that you usually need to start a project and save the whole thing as a Template, the next time you want to create a similar project you just open up the Template and all the settings are correct enabling you to get started quickly. The Quick Start dialog when opening SONAR includes a number of pre-made Templates and selecting one of these may suit your needs but a custom Template can be set up easily and quickly using just a few simple stages. Although you may not be ready yet to create your own Template files you can always refer back to this section as it is alongside the other 'groundwork' topics. As you get more used to working with SONAR you will develop your own procedures and then you'll probably want to make life easier for yourself with a few Templates.

Here's how to do it:

1. Create a New Project using the Quick Start dialog that appears when you open SONAR or select New from the File menu. Choose an existing Template that's close to what you need but don't worry too much as you'll modify it anyway.

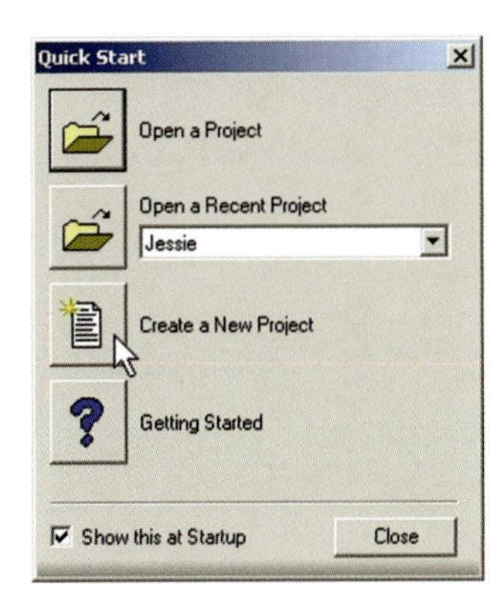

2. Arrange the Tracks, their type, order and name them to suit your requirements. You can also Arm Tracks for recording if you want to.

3. Set up any Inputs and Outputs that you normally use to match your soundcard (for instance eight channels to match an ADAT system).

4. Insert any effects and soft synths that you need.

5. Make any other adjustments such as tempo, which View (Console, Piano Roll, etc.) you want to start with.

6. Save the Project using File>Save As with a suitable name as a Cakewalk Template file (.cwt) in the SONAR Sample Content folder by selecting

Template as the Save as Type and choosing Template Files (.CWT,.TPL) in the Go To Folder field of the Save As dialog.

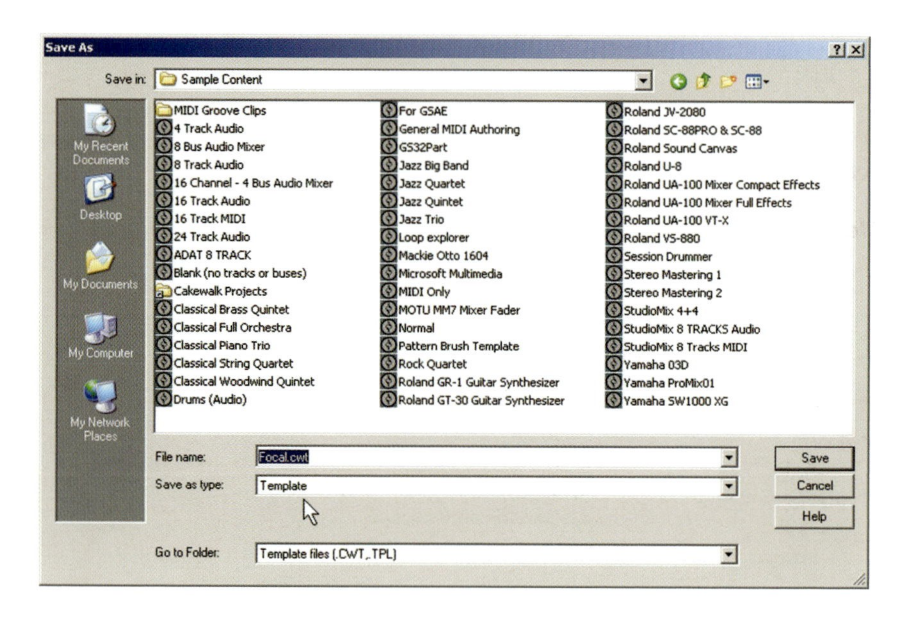

7 Next time you start a New Project File you'll find your Template among the list available.

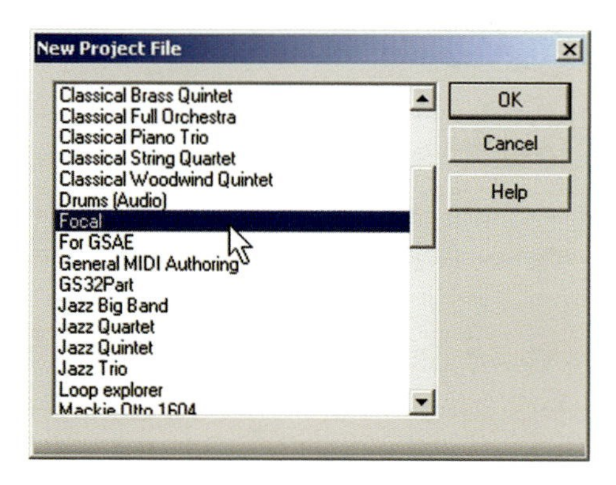

Template Tip

If you find that you're using the same Template all the time then you can save it under the name 'Normal' and overwrite the original file, this way it will be the default choice when starting a New Project.

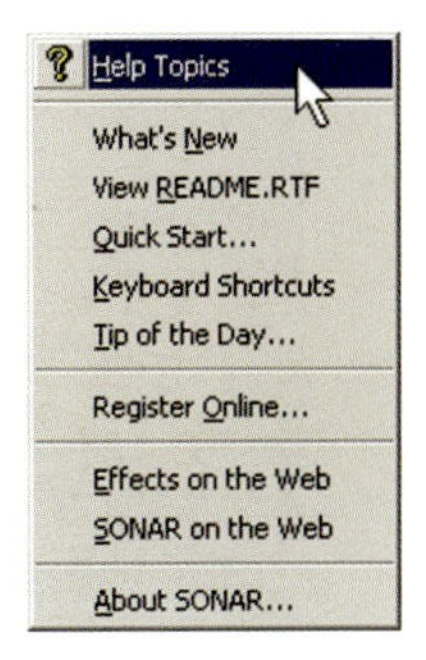

SONAR Basics: Things You Need to Know

Help

Context sensitive Help files are available at any time by pressing F1 on your keyboard or you may access the searchable Help system from the Help menu.

Views

A variety of specific Views are available within SONAR and they can be displayed simultaneously with other Views unless maximized. The style may be defined from the Window menu. If you have a dual-monitor screen setup then you can display more than one View by dragging a View to the second monitor while leaving the main View on the main monitor.

Track View – The main interface that you'll probably use is the Track View which consists of two areas or 'Panes' namely the Track Pane to the left and the Clips Pane to the right, separated by a moveable, vertical splitter bar. At the bottom of the Track View is an area that can be shown or hidden by clicking the Show/Hide Bus Pane button ( ) or dragging the horizontal splitter bar. The Buses appear in a similar fashion to the tracks above but do not contain any waveforms although they may contain other data such as Automation Envelopes. At the bottom of the Track Pane are four tabs which select the tools available within the Track Pane and the Bus Pane when visible, they are: All for all controls, Mix for mix specific controls such as Volume and Pan, FX for plug-in effect controls and Sends to Buses and finally I/O for Inputs and Outputs (All Mix FX I/O).

Tracks may be hidden either from their right-click menu (right-click on the Track number)

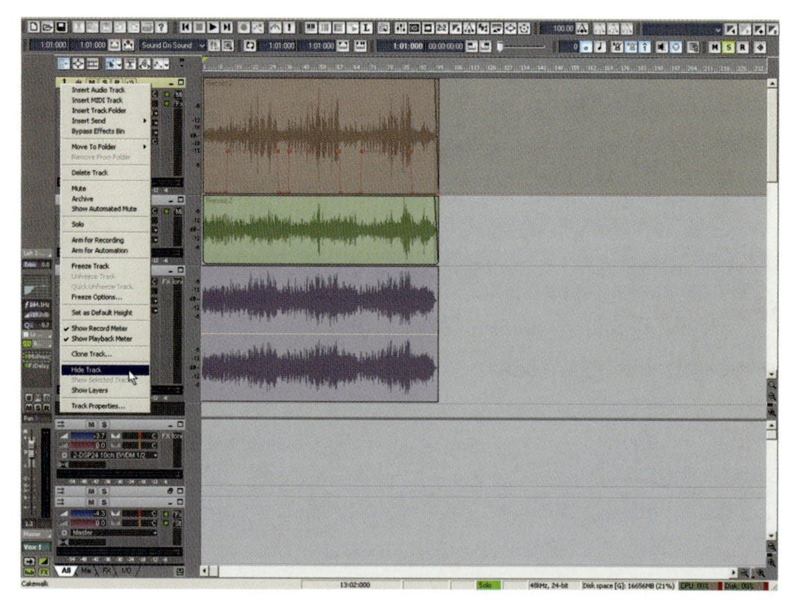

or using the Track Manager [M] to select or deselect them.

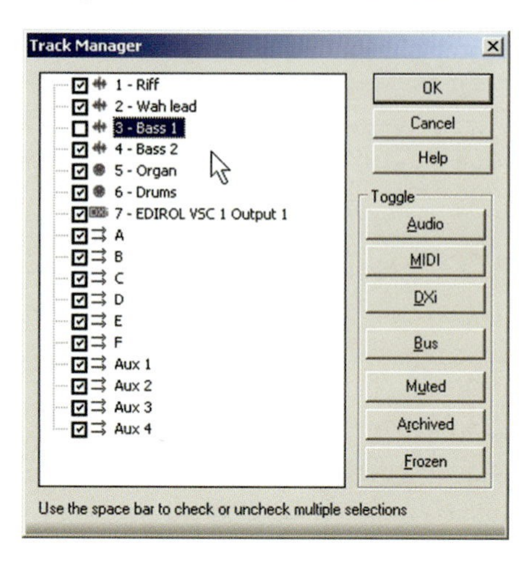

Console View – Apart from the Track View you may well use the Console View quite often (I know I do). It's based around a traditional mixer in its layout and consists of Track Strips for every Track in the Project, both audio and MIDI. You can open it by clicking on its button in the View Toolbar ( ) pressing Alt+3 or choosing it from the View menu. It contains your main hardware Output Buses and can contain other Buses defined by yourself whether for auxiliary effects or as Subgroups. Any Strip can be hidden by selecting Hide Bus or Track from its right-click menu or by deselection in the Track Manager [M].

All normal operations can be carried out here but sometimes the View is better, particularly when working with inserted effects as everything is easier to see.

Inspector Strip

To the left a Track/Bus Inspector can be displayed with a channel strip of the currently active Track; this can be turned on or off using the Show/Hide Inspector button ( ) or I key. At the lower end of the Inspector are a set of controls that enable tools specific to the channel to be viewed or hidden. Some of them such as the FX and EQ (FX and) buttons can be toggled to show several states. It also has a right-click menu which contains a 'Narrow Strip' option allowing you to gain a bit more screen real estate but keep the Inspector visible.

Toolbars

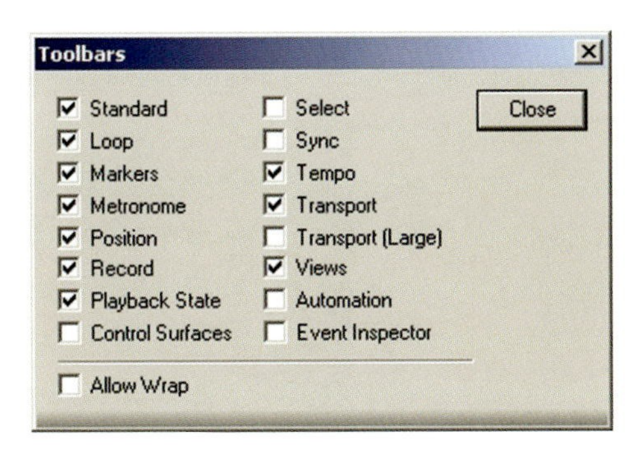

A wide variety of Toolbars are available via the View>Toolbars menu for the most commonly used commands and utilities (right-clicking on a Toolbar button also opens the Toolbar dialog). They can be docked at the top or bottom of the screen and arranged in any order, it's also possible to 'float' any Toolbar and move it to any position as desired. Toolbars without edit boxes can also be docked at the left or right of the screen.

Track Navigator

The Track Navigator is an overview of your Project and its Tracks in an easily visible format enabling you to see, and move to, sections that are currently off-screen. You can drag and resize the Navigator's Track Rectangle in order to see any relevant parts of your composition in the Clips Pane. You can show or hide this by using its button ( ) at the top of the Track Pane or by pressing D.

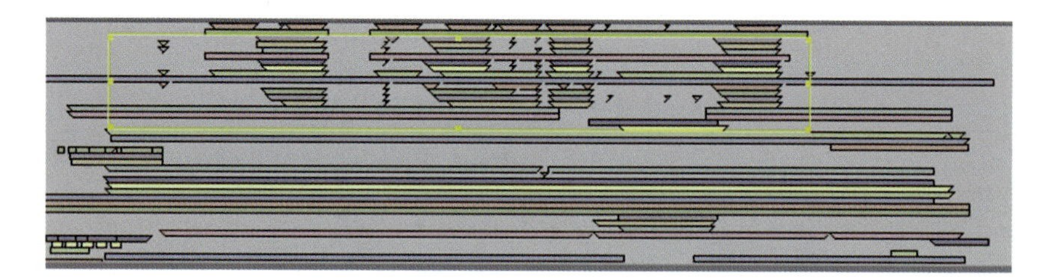

Track Tools

The most commonly used tools have options buttons positioned at the top of the Track Pane ( ). Clicking any button will turn it on or off and each one has a Tool Tip describing its use that becomes visible if the mouse cursor is paused over it, this also shows the keyboard shortcut key to select it in square brackets like this [T]. Some also have drop-down menus to adjust their parameters which are easily accessed by clicking the down arrow next to the button. If you can't see them all then just drag the vertical splitter bar to the right or click this button.

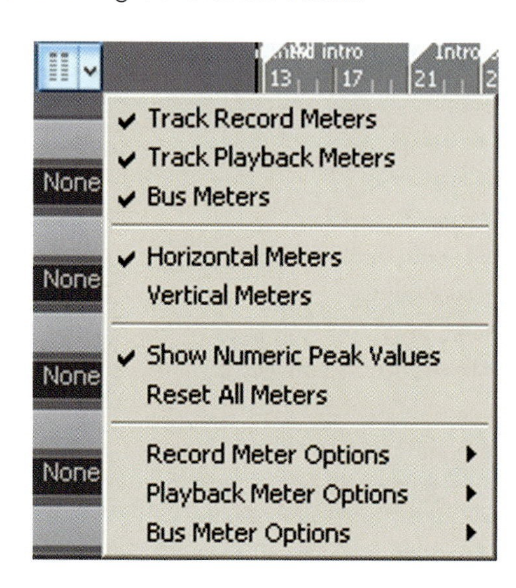

The three buttons on the left, simply: Show/Hide the Inspector, Navigator and Video Thumbnail Track ().

Next to them are the Select Tool (which is the one you'll probably use most for general selecting of Tracks, Time Ruler settings, etc.) and the Split Tool for making edits ().

The Scrub Tool () is used to drag over the Clips to make them play as it moves over them, a bit like winding tape over playback heads. You can drag forwards and backwards, useful for finding edit points, and dragging in the Time Ruler Scrubs all Tracks. Holding down the Alt key allows you to Scrub at unity gain.

The Envelope Tool is used for drawing and editing Automation Envelopes without accidentally dragging Clips ().

Smart Mute allows you to drag over a Clip and Mute its contents temporarily ().

The Zoom Tool enables controlled Zooming of a defined area; click and drag with this tool to lasso an area which will then be magnified to fill the Clips Pane (). Alternatively you can hold down the Z key temporarily then drag over a region to Zoom.

Snap To () is very important as it defines where any moved objects 'stick'. The available resolutions are set using its drop-down menu and range from Measure to Frames or it can be turned off completely. Note the Move To and Move By options, as this will become important when editing.

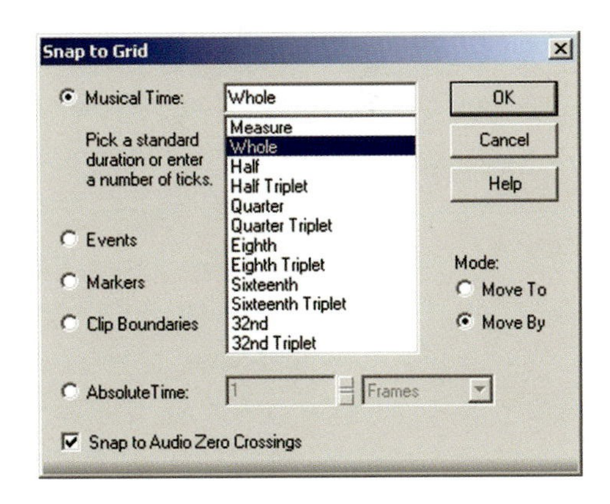

Automatic Crossfades can be invoked by activating this button (). When active, any overlapping audio Clips will have Crossfades applied, the curves used are as defined in its menu.

Meters can be shown or hidden with this button ( ) and its menu allows you to see them vertically or horizontally and also to set a wide range of parameters for them. Further meter options are available from the Options>Audio Meter Settings dialog.

Other Views

Views other than the Track View can be invoked using the Views Toolbar ( ) or View menu, some are automatically opened when a Clip is double-clicked, the defaults can be changed by right-clicking in the Clips Pane and using the View Options menu.

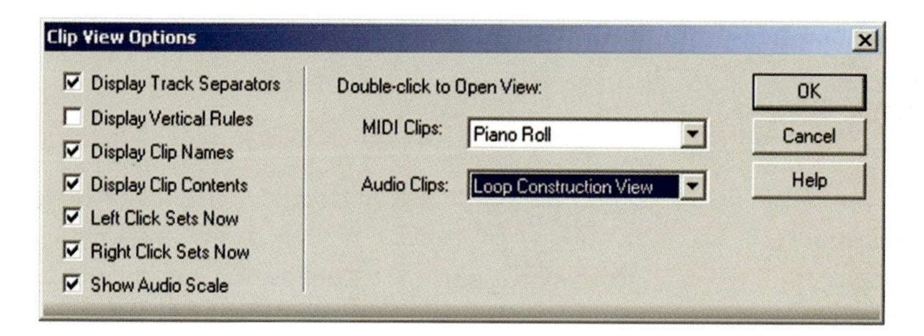

Each View has a specific purpose to enable specialized operations to be carried out either on MIDI or audio Clips and Tracks or the overall Project.

CHAPTER 3

RECORDING AUDIO

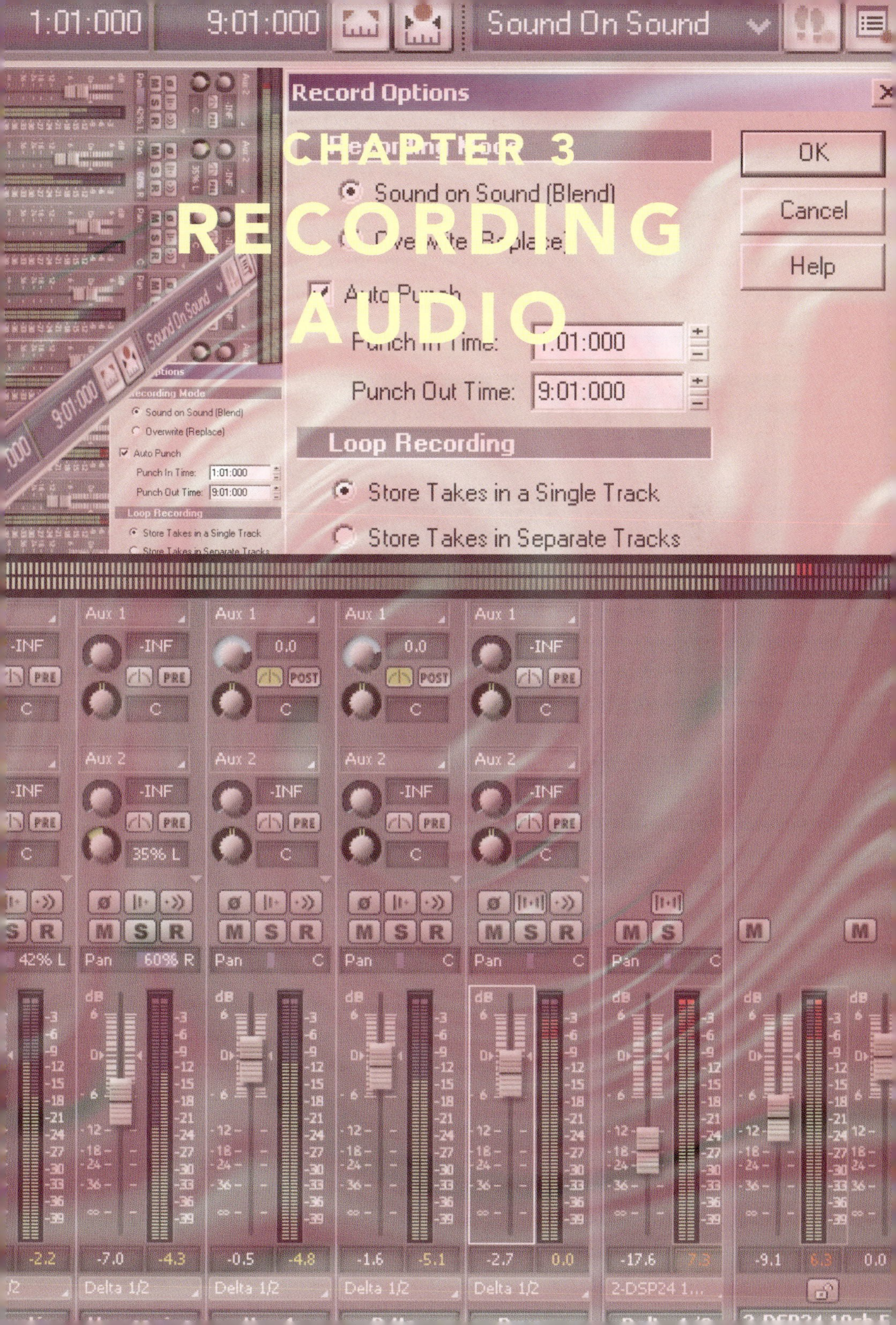

The obvious mainstay of any recording studio is audio, even if you write your music entirely using soft synths you will, at some stage, need to convert or record your creations as audio.

Audio can be the sole format that you use or can be brought into play just to record a sample that will be tweaked and contorted then used as the basis for a synthesizer or sampler patch but it is vital and everyone using SONAR should learn how to record, edit and process audio.

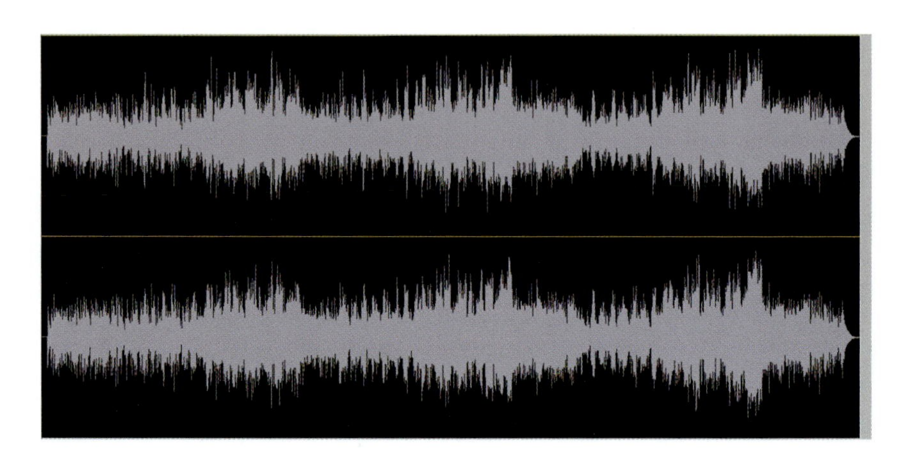

Bits and Hertz

If you're familiar with audio bit rates and sampling frequencies then you can skip this section but if you're not then read on.

Without going into too much detail the basic principles behind audio recording state that the higher sample and bit rate you use, the better the quality of the finished piece will be (that's in theory) but there is a trade-off; higher resolutions need more processing and more storage space. There is no room within this book to look at the many arguments both for and against but bear in mind that CDs are 16 bit at 44.1 kHz and they sound pretty good at that rate. Your hardware may dictate your limits to some extent but most will allow you to run at 24 bit (which I personally use) and usually above 44.1 kHz if you want to (I use 48 kHz). 24 bit is useful as it will allow you more 'headroom' when recording audio, that is, you won't hit the limit, or 'clip' point as easily, you'll know when you do as the meters will go into the red and you'll get a horrible distorted sound coming out of your speakers. Digital clipping isn't very forgiving like tape used to be and it sounds horrendous!

The settings are available in the Options>Audio menu and you should set Driver Bit Depth to 24 bit if your hardware supports rates over 16 bit. The File Bit Depth setting defines the resolution SONAR uses to store the audio files but remember that 24 bit uses much more space than 16 bit, and will use more resources.

Getting the Signal In

Your soundcard will to a great extent dictate your inputs but usually you'll have at least one stereo socket as a mini-jack, standard 1/4-inch jack or maybe phono sockets. Your sound source should be connected here but depending on its type you may need a plug adapter, pre-amp or mixer to obtain the correct signal level. Some soundcards have these built in. When recording you'll probably need to open your soundcard's mixer utility to ensure that you are recording the correct source and nothing else; some will have a setting to record what you are hearing which will record any sound that you are listening to such

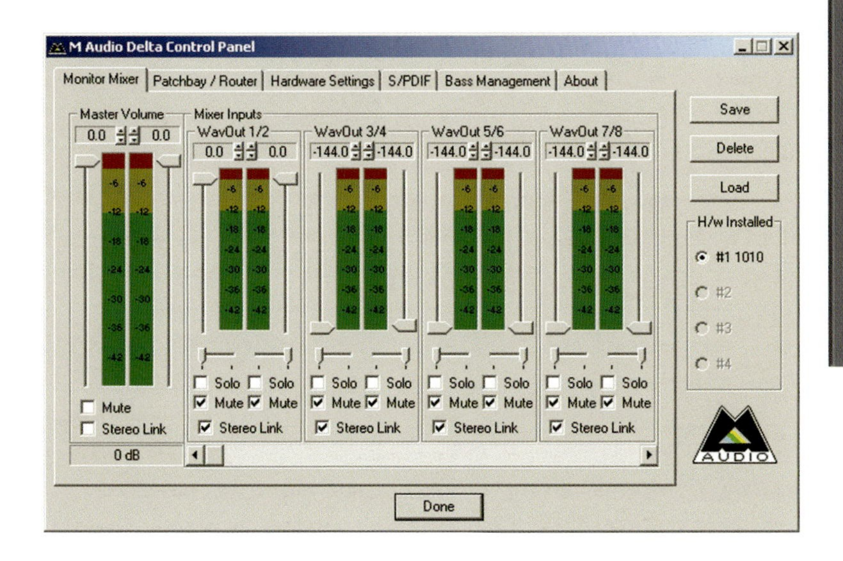

as the Metronome as well as the one you're trying to record. If in doubt then check the soundcard documentation, as this can be the cause of many frustrating problems yet it is easy to sort out if you take a little time to understand the mixer controls. If you aren't sure then run a few test recordings and keep a note of the settings, you'll soon find out what works and this will also help you understand the workings of your system.

Importing Audio

Audio files can also be imported into SONAR in a variety of ways from any available drive on your system such as a CD. To Import audio you can use File>Import>Audio then browse to the file that you want to use. Alternatively you can use the Loop Explorer View which offers a little more flexibility by clicking its icon ( ) or selecting it from the View menu. The principle is basically the same as the Import dialog but we'll be looking at it in more depth later in the Editing Audio chapter when we look at Groove Clips.

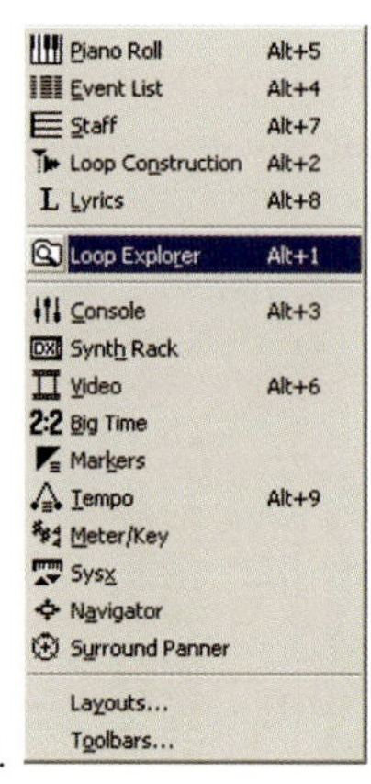

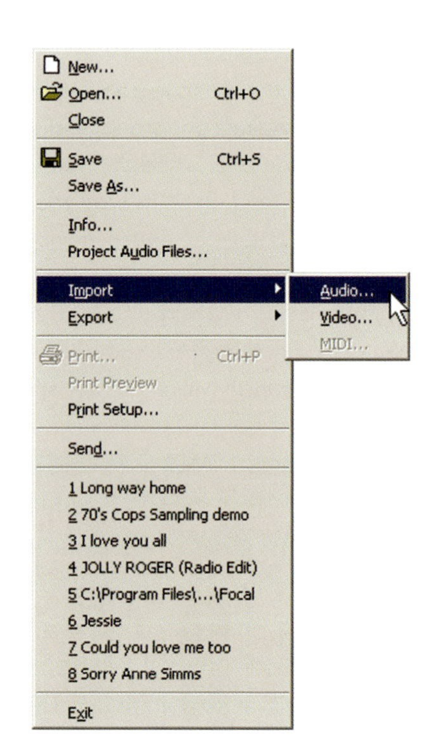

Input Monitoring

If you're not using an external mixing console or a soundcard that has direct monitoring facilities you may have difficulty hearing what you're trying to record along with what you're playing back. SONAR has an Input Monitoring facility that will allow you to monitor any audio tracks simply by clicking their Input Echo button ( ). If you hear a big delay between what you play and when you hear it back then you may need to change your latency in the Options>Audio dialog.

If you do this then turn your speakers down first in case you create a feedback loop! If necessary you can click the Audio Engine button ( ) in the Transport Toolbar to turn off all audio quickly.

Input Monitoring can also be used to hear inserted effects in real-time as you play or sing through them, they aren't recorded though so you can change them later or apply them destructively if you wish.

Setting the Levels

The main aim when recording audio is to get a high-quality signal into the PC which also means a high level. This will depend to some extent on your external devices as well as your soundcard mixer but always try and obtain a strong signal without going too high and distorting (or clipping) it.

To check the incoming level in SONAR do the following.

1. Open a Project or create a New Project. You can use the Quick Start menu that appears when SONAR starts, select New from the File menu for a new Project or Open to open an existing one.

2. Select an Audio Track or insert one using the Insert menu and selecting Audio Track.

3. Expand the Track's controls by clicking the Restore Strip Size button or dragging it open ( ).

4. Make sure you have the All or I/O tab selected at the bottom of the Tracks Pane (All Mix FX I/O).

5 In the Input (I) field click the arrow and choose the Input that you want to use. The options will depend on your available soundcard Inputs ( ).

6 Click the Track's 'R' button to Arm it ( ).

7 The Meter will now display the incoming signal level providing the rest of your signal chain is OK.

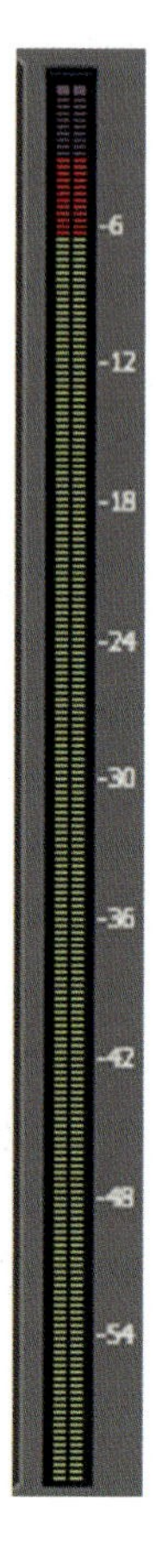

8 Play the sound source as it will be played on the recording and adjust the level coming in to peak just inside the red section of the meter or just below it, around the 6 (−6 dB) mark. You can play any other Tracks while doing this if you like.

While we're looking at meters it's worth checking out the Meter Options by clicking the arrow on this ( ) button, I prefer to use the Horizontal Meters setting and sometimes the Hold Peaks and Lock Peaks options available from the submenus can be useful when mixing a lot of tracks so that you can see what each one is maxing out at. You can also adjust many other meter settings to suit your own methods and match any existing equipment.

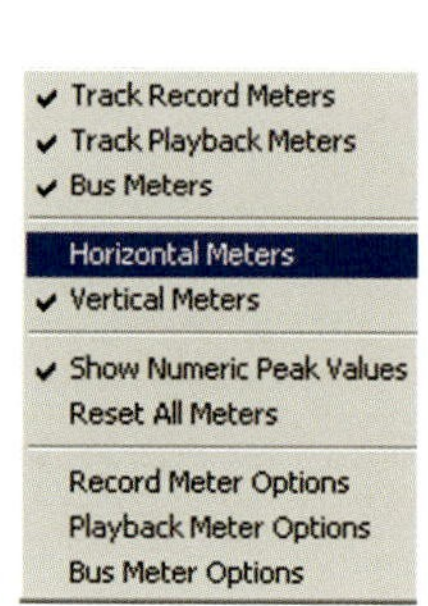

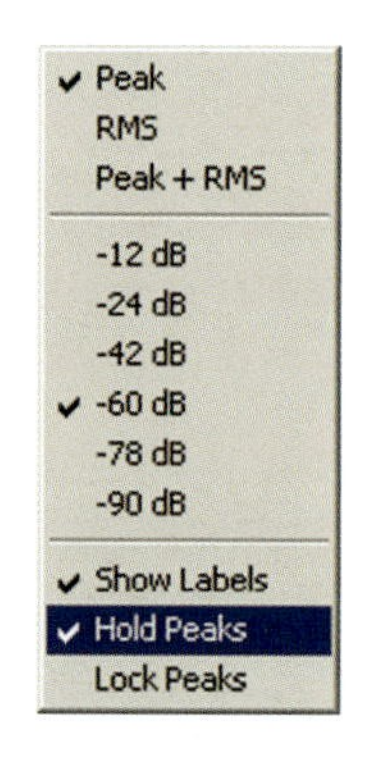

Before You Start Recording

Before commencing with your recording you'll probably want to set the Tempo and Metronome so that you have something to play to. The Tempo is easily set using the Tempo Toolbar (View>Toolbars tick Tempo) by clicking on the current BPM number and changing it using the + and − controls or typing a figure in, you can also click the adjacent Insert Tempo button ( ). If you're not quite sure of the Tempo but have it in your head then click the ( ) button then tap your Spacebar at the correct tempo then click OK, this will automatically adjust to suit the Tempo you tapped out.

Make sure that Change the Most Recent Tempo is selected unless you want to insert a tempo change. Tempo changes can be inserted anywhere in a Project, just put the Song Position Pointer where you want to insert it and then adjust the Tempo in the Insert>Tempo Change dialog (the same one that we just accessed from the Tempo Toolbar) making sure that Insert a New Tempo is selected. Note that if you change a Tempo after audio has been recorded or imported it won't follow the change unless it's a Groove Clip [See Groove Clips] but MIDI Tracks will.

The Metronome

The Metronome is set from the Options>Project>Metronome menu

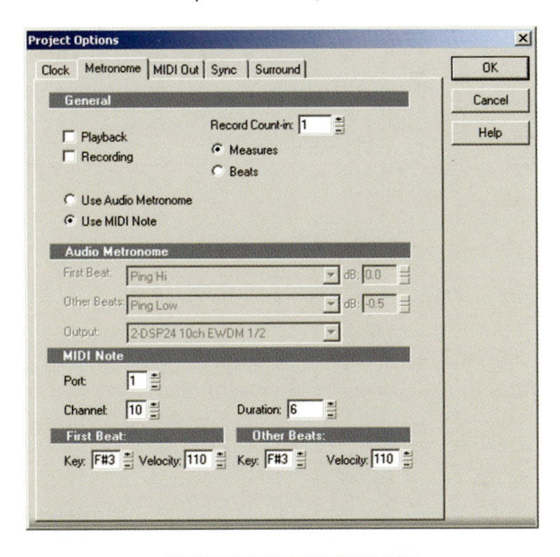

(see Chapter 1) or its Toolbar ( ) and can use either a

connected MIDI device or play audio sounds, you can also add your own sounds to it by putting them into the Metronome folder. You can specify a different sound for the First Beat than that used by the Other Beats and change their relative volumes with the dB-offset controls. It's useful to use sounds that are unlike the one being recorded so that the performer can hear them easily, if you use hi-hats for a drummer they may not be heard very well against the drummer's own hi-hat sound!

Another Metronome option you may find useful is its Record Count-In feature which can be used to play the Metronome for a specified time before the recording starts.

Mono Recording

When recording a mono source such as a guitar or vocal you need to select one side of your soundcard's input such as its Left Input which is usually the first one in the list. If using a jack plug input this will work when a mono plug is used in a stereo socket. If using a hardware mixer between the source and the soundcard then it's preferable to pan the channel on the mixer to the relevant side for better signal integrity.

When everything is ready and your signal is good just click the Record button () or hit the 'R' key on your keyboard to start recording and click stop or hit your Spacebar to stop. You will see the waveform appearing as you record but remember that you can only record on Tracks that you have Armed ().

Stereo Recording

When recording a stereo source such as a stereo synth or a pair of drum overhead microphones you have the choice of recording two separate mono signals and panning them to their respective sides using the Pan slider () or recording as a dedicated stereo track. To do this just select a stereo input from the list available which will appear below the Left and Right names of the same input socket.

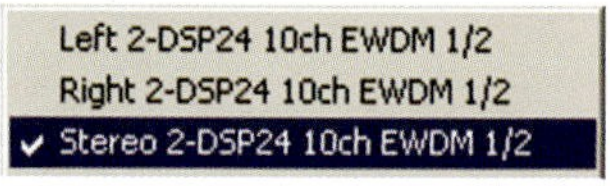

Left 2-DSP24 10ch EWDM 1/2
Right 2-DSP24 10ch EWDM 1/2
✔ Stereo 2-DSP24 10ch EWDM 1/2

Make sure that the relevant tracks are Armed, check that the signal is good and then click the Record button or press 'R'.

Recording from Multiple Sources and Phase Correction

Often it's necessary to record from multiple sources at once such as a drum kit, a live surround recording or when several musicians are playing a piece together such as a quartet. In this case you may mix the sounds into stereo before they enter the soundcard using a hardware mixing console but you could also use a multiple input soundcard or interface and bring them all in at the same time onto separate tracks. To do this you'll need to configure the inputs using the devices mixer utility and insert the relevant audio tracks in SONAR; you can use mono, stereo or a combination of both. They all need to be Armed and their Inputs set to correspond to the incoming signals, then each level should be set correctly while all the sources are active, that is, being played. Once everything's set the recording is started from the Record button or 'R' key.

If you plan to use this method regularly then it's an ideal candidate for creating a Template.

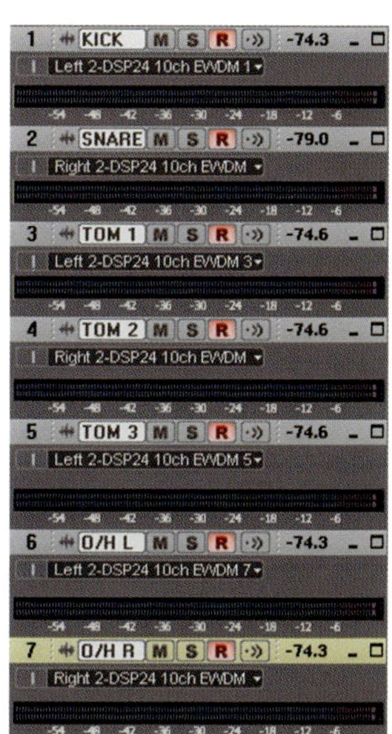

Tracks Routed to Multiple Inputs, Armed and Ready to Record

Whenever multiple sources are recorded there's a possibility of 'phase cancellation' which, without going into too much detail is caused by the peak of one signal coinciding with the trough of another signal coming from the same origin (e.g. a violin with two separate mics recording it) and cancelling out, causing a drop in level. This can be quite severe and result in serious level drop or distortion but can be cured by moving one of the mics causing the problem or moving the recorded signal back 'in phase'.

SONAR has a Phase Inversion switch ( ) on each audio channel which shifts the phase by 180 degrees and can be used to check and often cure phase cancellation, just click it on one of the affected audio tracks and listen (it will look like this (Inv.)) if the sound gets louder then the audio was out of phase and you can leave the button switched on to correct it. It is also possible to zoom in to the waveforms and manually adjust the phase by moving one of the tracks until the waveforms match but the switch will usually correct the problem.

How to Make a Basic Audio Recording

You'll need to set up your Tempo and Metronome first if you're going to use it or make sure it's turned off if you're not. The performer will probably need to have a pair of headphones on with the output of your soundcard going into them as well as their own signal. To do this you may need to use Input Monitoring mentioned earlier or set up a separate monitor mix from your hardware sound mixer. The tutorial below is for a basic mono recording but if you're making a stereo or multiple source recording just perform these steps for each source:

1. Attach your source (in this case a microphone) to the appropriate soundcard input.
2. Open SONAR and start a New Project, we're going to use the Normal Template.

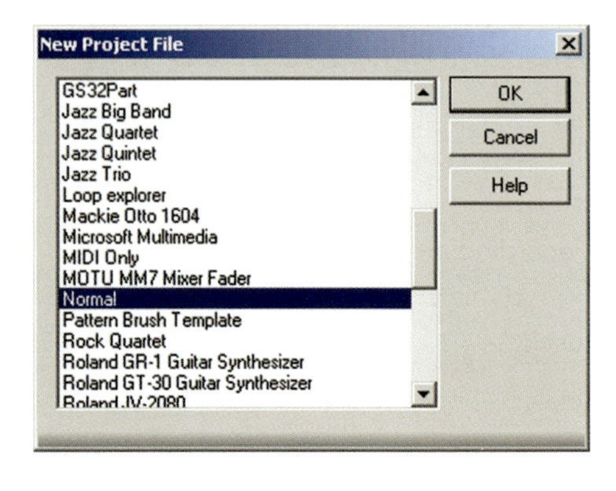

3. Arrange the Track View so that you can easily work with the first audio channel and set it's Input to match the soundcard input you are using.
4. Press the (R) button on the Track you are using to Arm it for recording (it will turn red).

5 The performer should now play or sing as normal while you open your soundcard's mixer (or use your hardware) and set the incoming level until it peaks just below the red section in the SONAR Track Meter on the loudest parts.

6 When all is well you can click the Record button or press 'R' on your keyboard to commence recording.

7 When the recording is finished click the Stop button or press your Spacebar.

8 Don't forget to Save! ( ).

If your recording is unsuccessful you can Undo it using the Undo button ( ) Edit>Undo or by pressing Ctrl+Z. It's always worth having a listen back to any parts that weren't quite right as it's often possible to make edits such as copying and pasting a good chorus in place of a bad one.

Any takes that you want to keep but don't need right away are best Archived by right-clicking on the Track's number in the Track Pane and selecting Archive from the menu. This effectively turns off the Track conserving resources.

Recording Multiple Takes Easily

Sometimes, particularly if you're recording by yourself, you may want to keep playing until you achieve a take that you're happy with. Usually this means that you have to keep stopping, rewinding and then starting the recording again but SONAR has an option that gets round this problem very neatly and here's how to use it . . .

First you need to set up a suitable amount of looping playback to work with, this has to include enough time for you to get into the track (say two bars) and a little time to get back out including any time for sustained notes that you may be playing.

1 Click and drag in the Time Ruler from the start to the end of the section that you want to use, the Ruler will darken for the selection ().

2 Make sure that you have the Loop Toolbar open (View>Toolbars>Loop) (1:01:000 1:01:000) and click the Set Loop to Selection button ( ).

3 Now play the track back and it will loop around the section that you just defined. If it isn't quite right just click and drag one of the yellow markers to the correct position ( ).

4 When you're happy with the selection insert an Audio Track (right-click in the Track Pane and choose Insert Audio Track), set it's Input and Output as desired then Arm it.

5 Go to the Transport menu and click Record Options then under the Loop Options heading check the Store Takes in Separate Tracks option.

6 Close the dialog box and you can now start the recording with the Record button or by pressing 'R'. The song will keep looping around your chosen points allowing you to keep playing until you get a take that you're happy with.

7 Stop the recording and you'll see that each take is stored in a newly created separate track ready for you to work with.

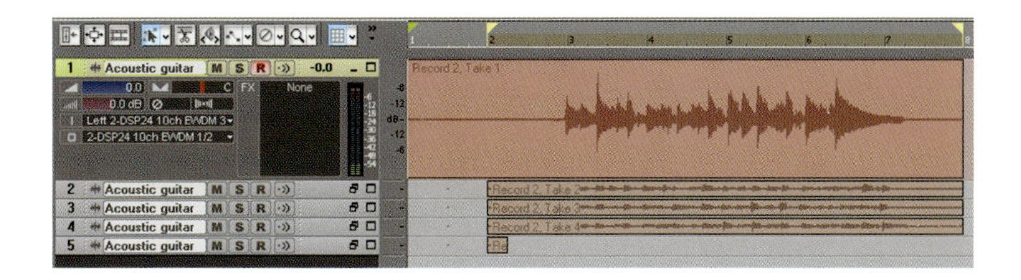

8 Each new Clip will be named after the recording number and the take number so these are from the second recording and are Takes 2, 3 and 4.

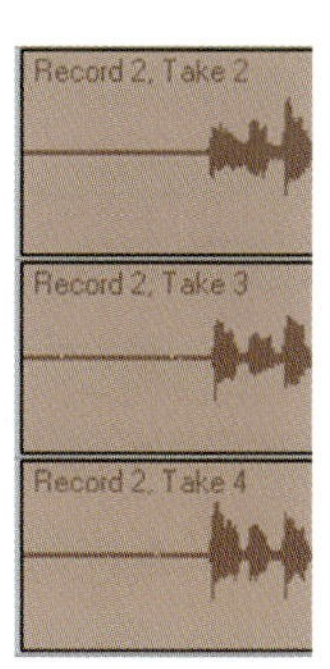

9 It's also possible to store more than one take in the same track, useful for compiling them later and conserving space. To do this you need to right-click on the Track (in either Pane) and select Show Layers. You can then choose to Store Takes in a Single Track from the Record Options or drag takes onto a Track to group them together.

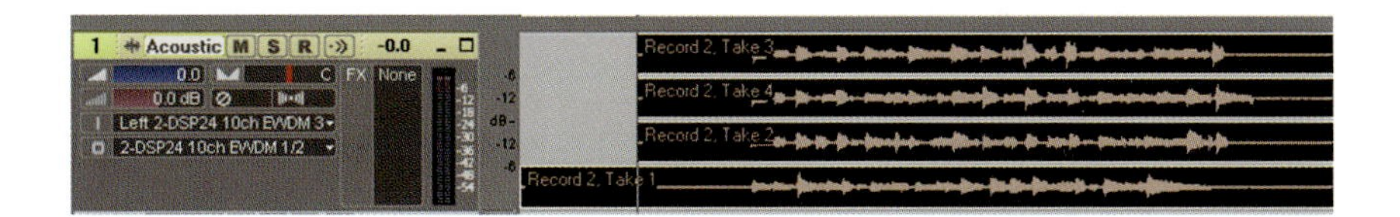

Saving and Exporting Audio

You'll probably be using most of your audio takes as part of a larger Project within SONAR but when you've mixed down to a final stereo track or if you want to export a sample for instance or take a few tracks into another package or studio for processing you'll need to Export them from SONAR. The quickest way is to just drag the Clip from the Track View onto your desktop, a folder or other suitable location but for a more refined approach use the File>Export>Audio dialog.

Here you can browse to a location on your system and specify the exact file format, sample rate, bit depth, etc., and also choose to apply any effects and automation during the process. The Preset box enables you to save your most often used combinations for fast recall.

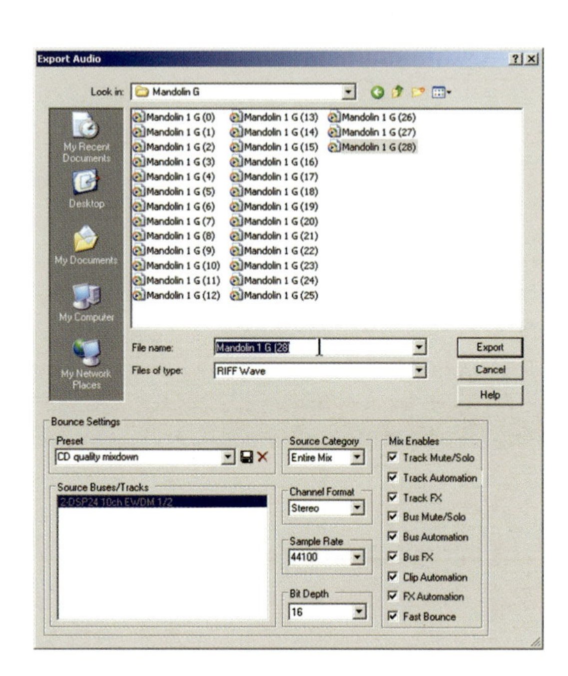

Audio Tips

If you have a multiple input system and do a lot of simultaneous multi-track recording then it's worthwhile creating a Template with the audio tracks already configured for each Input and Output, and Armed ready to go for recording. You can also have the Tempo and Metronome at your usual settings enabling a fast start for new projects.

Whenever you have audio tracks that you don't need to use immediately it's a good idea to turn them off completely to conserve system resources. Muting will just stop them playing back audibly but if you right-click on the track's number you can select Archive which will turn the track completely off. To unarchive just do the same right-click and untick the Archive option.

You may like to use the larger Playback Toolbar which is available by going to the View>Toolbars menu and ticking Transport (large) or by pressing F4. This Toolbar also has several other useful controls at hand including the loop, record and Metronome options.

To speed up your 'takes' don't set up a new track each time you record just drag the recorded Clip down to an empty space where it will create a new Track then either press the new track's 'M' button to Mute it or right-click on it's Track number and select Archive from the menu.

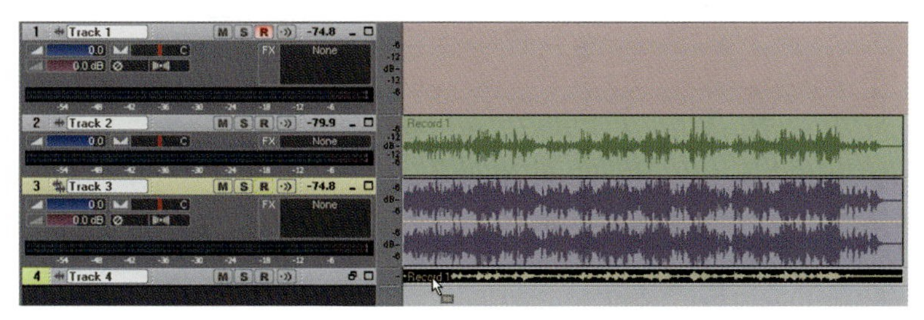

The Track that you were recording on is still configured and Armed ready to do another take so you can just click the Rewind button ( ) or press the 'W' key to go back to the start and then begin the next recording.

You can also select multiple Clips by holding the Ctrl key and clicking each one then drag them all down to an empty space together; they will create new Tracks for themselves the same as a single Clip but you will need to Mute or Archive them all as before, if they're all selected then you can use the right-click menu to do this to them all at the same time.

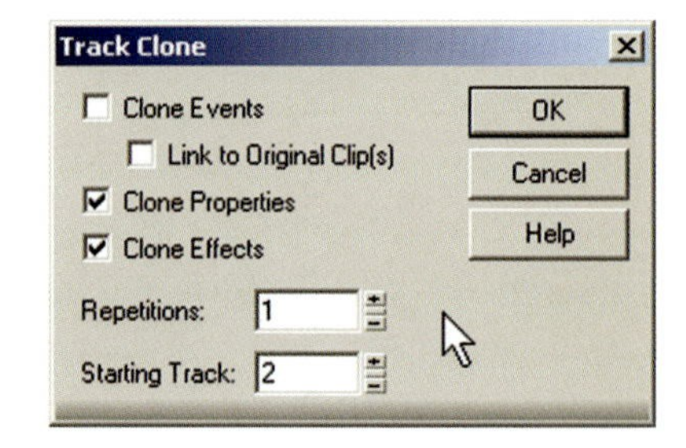

You can create a Clone of any selected Track by choosing Clone from the Track menu. Its dialog box provides options for the Clone.

CHAPTER 4

RECORDING MIDI

What's with All These Ports and Channels, and Things? (A Basic Overview of MIDI)

If you've never encountered MIDI before or have just a smattering of knowledge then it's worth learning a bit more about it even if you avoid getting too involved with it as it can also be employed when dealing with audio for things such as Remote Control and Automation even if you never use a MIDI instrument or soft synth.

The word MIDI is derived from Musical Instrument Digital Interface, which is exactly what it is; a system that allows two pieces of compatible equipment (hardware or software) to communicate via a common interface. Although it has a universal format, General MIDI or GM for short, there are also variations such as Roland's GS and Yamaha's XG format which expand somewhat on the original format. The most common application is to use one piece of equipment to control another; for example, a keyboard can play sounds from another keyboard, sampler or sound module. The standard connection is a five-pin DIN plug and socket arrangement but the connections are not the same as the old hi-fi DIN specification so although these cables will work it's a good idea to use only MIDI-specific cables.

The basic MIDI system controls a range of pre-defined parameters such as note messages, velocity, volume, pan, modulation, etc. within a fixed range of 128 steps. A silence would have a value of 0 while full volume would be 127 (some systems don't use 0 so silence would be 1 and full would be 128). When a key is pressed it will send a message that may be something like C4 on at velocity 114, where some controllers are constantly being sent such as volume, modulation, etc. so the receiving equipment already knows these parameters; these are called Continuous Controllers or CC for short.

The MIDI Tree

MIDI signal paths can seem complex but if we view them as a tree then they are easier to visualize.

We can start with a MIDI Port which has 16 Channels viewed as 16 branches all from the same tree.

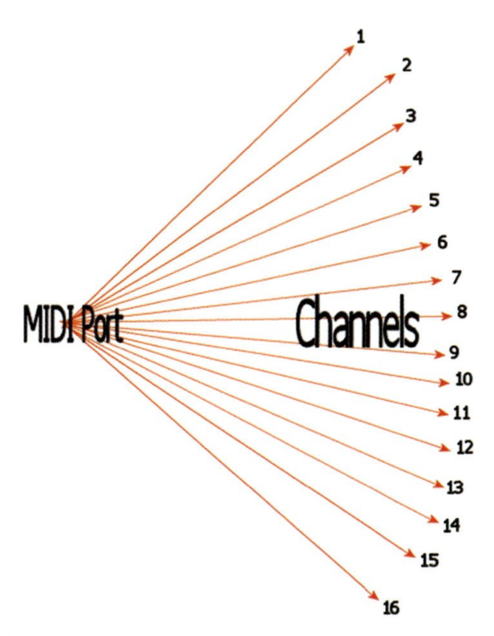

The Port can send and receive these signals so a duplicate of them can be visualized returning to the tree but down the same path.

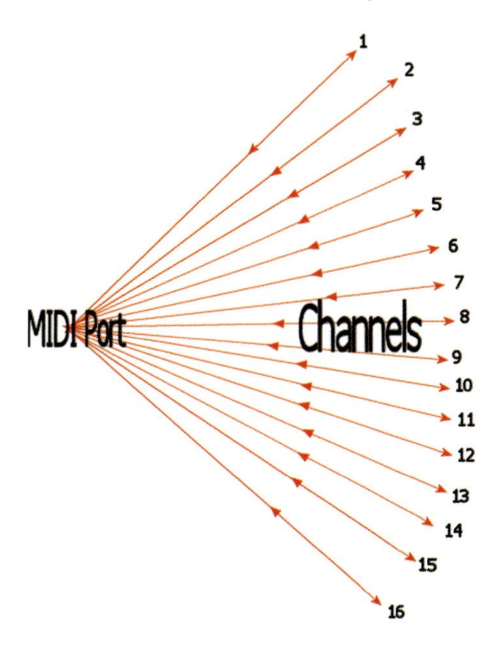

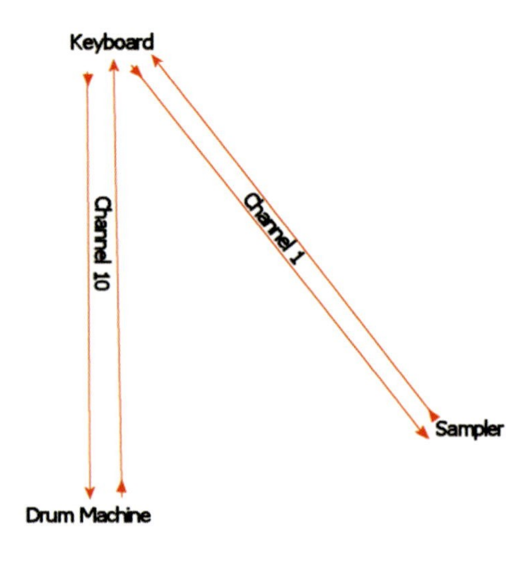

Each channel allows data in either direction and each channel has a number, much like a house address, which allows data to be sent to a specific location, for instance: Channel 1 has a sampler attached to it with piano sounds loaded and is set to receive only on Channel 1, a keyboard is set to transmit on Channel 1 so if the sampler is connected to the keyboard and the keys struck then the piano sounds will be triggered and heard from the sampler's outputs. If the keyboard was transmitting on Channel 10 only, the sampler wouldn't play at all BUT if a drum machine was connected to Channel 10 it would play its sounds.

There is also a setting called Omni Mode which will allow equipment to receive any incoming channel so if the sampler and drum machine were set to this they would both play when the keys were struck.

Put simply MIDI data contains the performance and not the sounds, they are produced by the soft synth or hardware sound module when it receives this data.

This is a very simple view of MIDI and there are a lot of other features, many of them specific to manufacturers and instruments that I don't have the space to cover but if you're interested then there are many resources available both in print and on the Internet.

Banks and Patches

MIDI sound modules and soft synths will often have Banks which can be selected. A Bank is a collection of sounds or Patches usually with common characteristics such as a Bank of organ sounds. Probably the best known will be the General MIDI Bank which is an eclectic mix of 'bread and butter' sounds ranging from pianos through orchestral instruments and synths to percussion and sound effects.

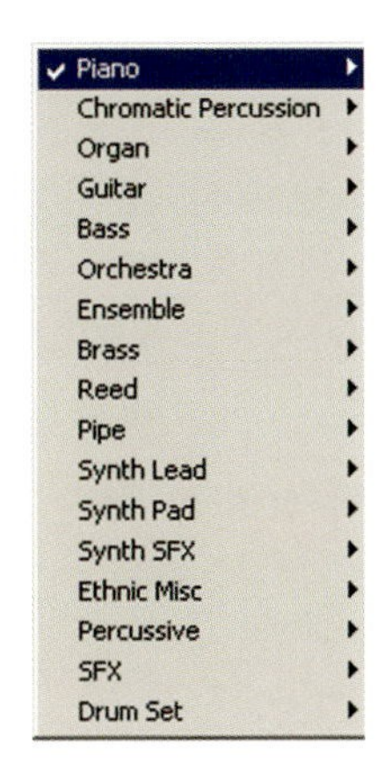

These are the Banks available from the Cakewalk TTS-1 synth included with SONAR.

Patches are the single sounds or instruments available either in Banks or on their own, for example: A Yamaha drum kit may be selectable from within a Bank of drum kits but it may also be a single preset on a synth.

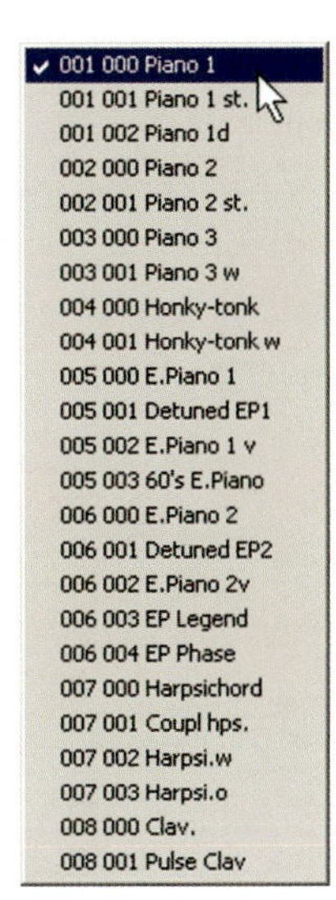

These are the Piano Patches available from the Cakewalk TTS-1 synth included with SONAR.

Painless MIDI Signals

In order to enable your equipment to communicate you will have to have a suitable MIDI interface in your computer system, this may be built into a soundcard or be a separate device such as a USB MIDI interface. Your MIDI keyboard, drum kit or whatever will need to be attached to it correctly using a MIDI cable and any relevant drivers must be installed.

The cables must be connected to an Output on one end and an Input on the other; so to play a controller keyboard (Transmit) the cable is plugged into its Output while the other end will go to the computer MIDI Input.

The MIDI interface will appear in SONAR's Options>MIDI Devices list of Inputs and must be selected by clicking on it so that it is highlighted and thus active.

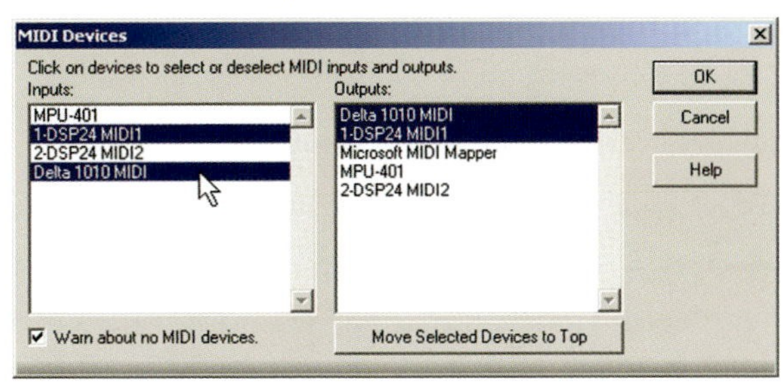

When SONAR is running a small icon ( ) appears in the system tray to show MIDI activity so if you're all connected up properly and you play your MIDI instrument the Input light will flash ( ).

How to Get a Sound Out with a Hardware MIDI Module

In order to hear a sound you will have to provide a source for your MIDI device to trigger which can be a hardware device attached to a MIDI Output Port or a soft synth running in SONAR. To do this with a connected hardware MIDI device such as a sound module . . .

1. Make sure your MIDI controller (we'll assume it's a keyboard for now) is connected up properly and that its signals are reaching SONAR so that the MIDI activity meter is flashing.

2. Ensure that your MIDI interface is installed correctly and is selected in the Options>MIDI Devices>Input section of SONAR.
3. Connect your module to the MIDI interface using a proper MIDI cable connected from the interface's MIDI Output to the module's MIDI Input.
4. Power up all of the equipment.
5. Check which Channel the MIDI module is set to receive on.
6. Open a SONAR Project and Insert a MIDI Track by right-clicking in the Tracks Pane and choosing Insert MIDI Track or use a template that contains one.

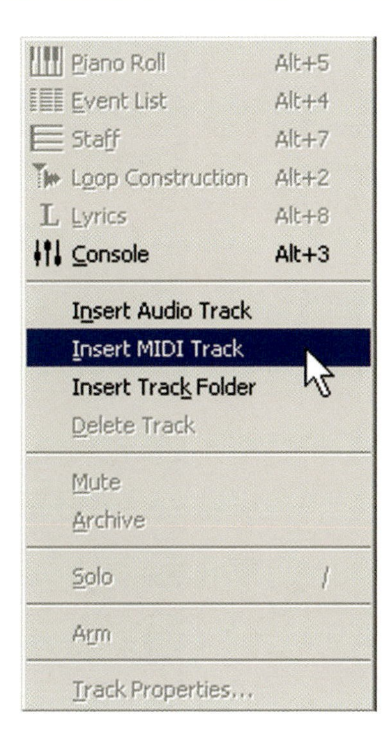

7 Set the Track's Input to the channel your keyboard is transmitting on or just use Omni Mode.

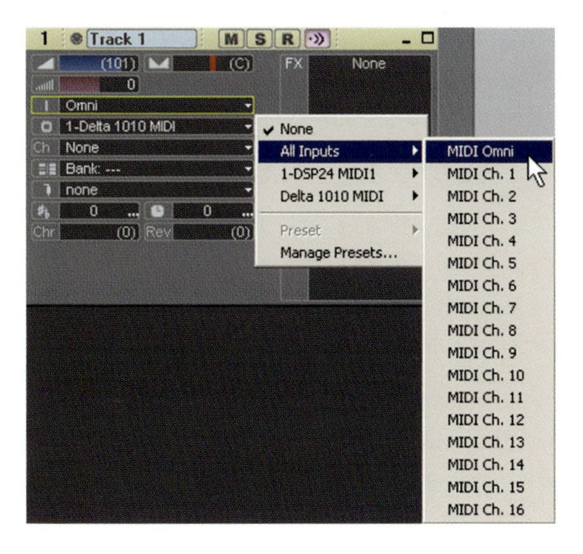

8 Set the Track's Output to the MIDI interface's Port that the module is attached to (there may be more than one).

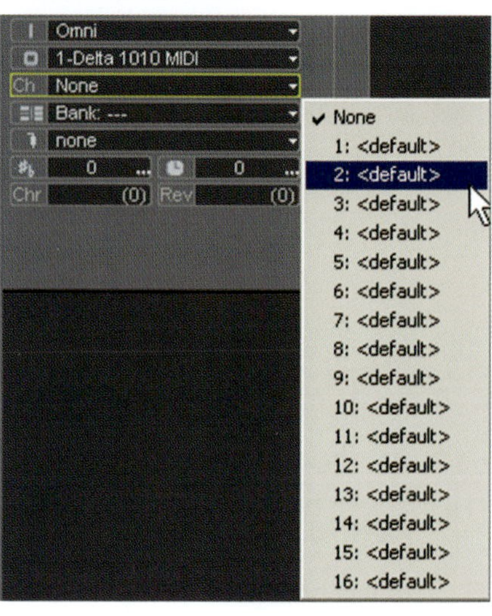

9 Set the Channel to the one which the module is set to receive on.

10 Play!

Providing that you have all other connections set up correctly, such as the module to an amp and speakers, you should hear the module's sounds played when you strike the keyboard's keys. If not then work your way through the instructions again and check that your signal path is correct at all stages, your module will probably show MIDI activity if it is receiving any data.

This method can also be used to route existing MIDI Tracks, such as a MIDI file, out to a connected module.

How to Get a Sound Out with a Soft Synth

We'll assume that you have a MIDI keyboard attached correctly as in the last tutorial but this time you want to make use of a soft synth's sounds:

1 Load up a new SONAR Project, a blank will do or delete any Tracks so that it is empty (right-click on the Track's number and choose Delete Track from the menu).

2 Click the Synth Rack button ( ) in the View Toolbar or choose Synth Rack from the View menu.

3 Click the Insert button ( ) in the Synth Rack then choose DXi Synth>Cakewalk TTS-1.

4 In the Insert DXi Synth Options dialog make sure that MIDI Source Track, First Synth Output (Audio) and Synth Property Page are checked then hit OK.

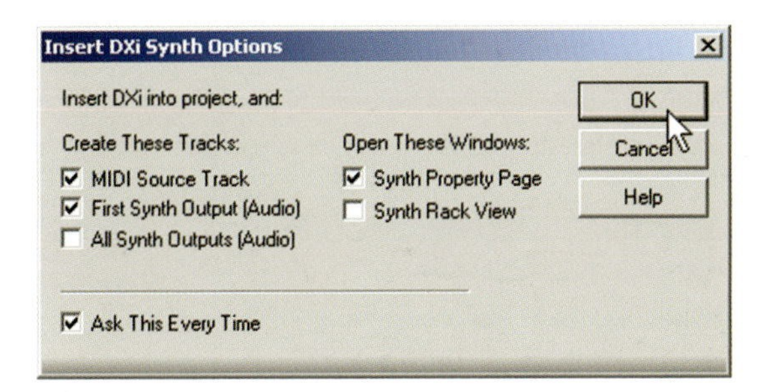

5 The synth will now appear and it will also have inserted a pair of Tracks, one is its MIDI Input and the other is its audio Output. Arrange the screen so that you can see everything clearly.

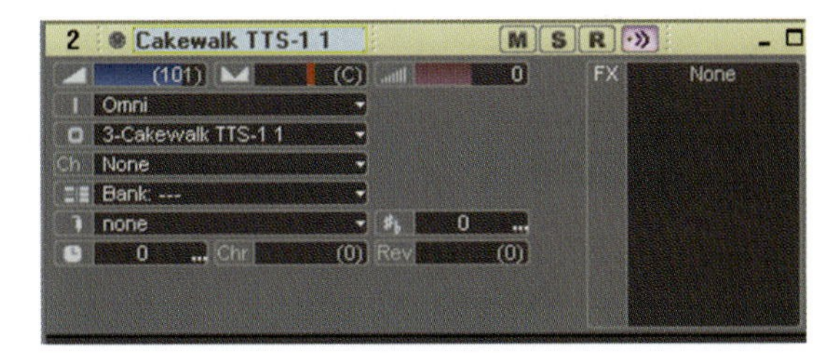

MIDI input

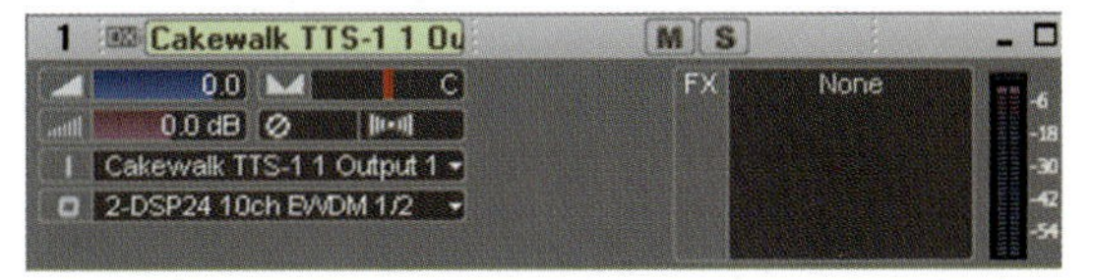

Audio output

6 The MIDI Input should be set to receive your keyboard's transmission Channel or Omni Mode to receive any incoming MIDI signal.

7 As the MIDI Output is already routed to the synth's Input you should now be able to play your keyboard and hear the sound from the soft synth; make sure that the synth MIDI Track is selected as the current Track (click on its Track number to highlight it). You should see the audio meter displaying the output signal.

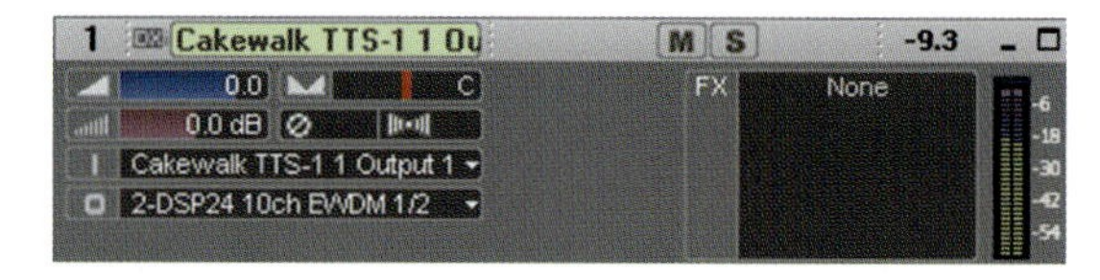

If you experience any 'lag' or delay between striking a key and the sound being heard try lowering the Latency setting by using the Mixing Latency slider in Options>Audio>General. Drag it to the left-side Fast setting then close the dialog, you should now have much less delay, the exact amount will depend on your interface's drivers. If you experience 'glitches' or 'dropouts' in your sounds then you may have to move the slider back towards Safe a little bit.

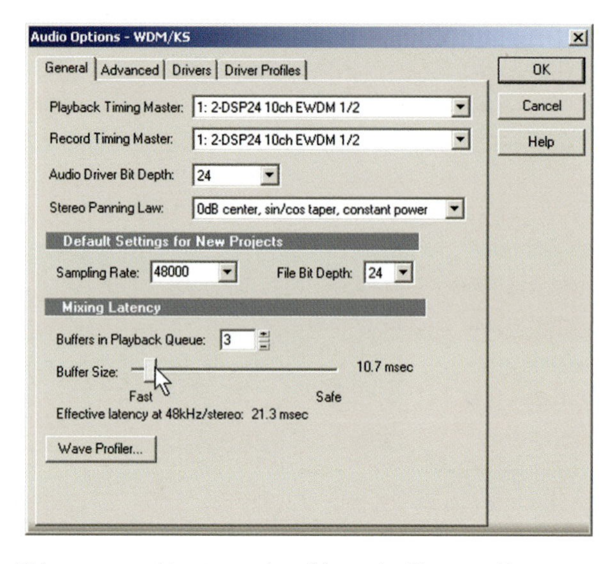

Soft synths will be covered in more detail later in Chapter 7.

Setting Up a MIDI Track and Recording onto It

Once a MIDI Track is inserted and configured to receive data correctly it can be recorded onto (or into if you prefer) in much the same way as an audio Track.

It isn't necessary to set the Input level in the same way although you can adjust Volume, Pan and Trim, etc. if you want to but usually this won't be necessary if your incoming signal is OK as there are many ways to tweak it later and clipping isn't a problem with MIDI so the chances of signal overload are virtually nil.

Arming a MIDI Track for recording is achieved by clicking the (R) button but note that no metering will be seen on the MIDI Track as none is required.

Pressing the Record button ( ) (turns red when recording) or 'R' key will start the recording process and when notes are played you'll see them appear in the Clips Pane as the song progresses.

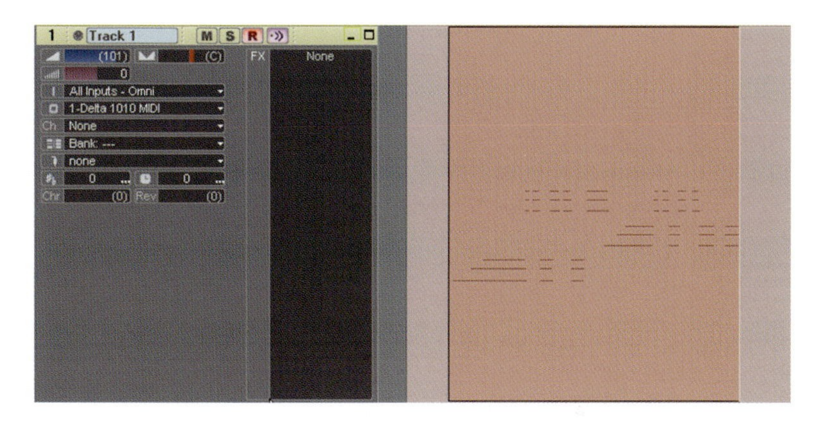

When stopped the MIDI Clip will be visible along with a visual representation of the notes recorded that may be edited later (see Chapter 6).

Inputting Notes in the Piano Roll View

The Piano Roll View is invoked by clicking its button ( ) in the View Toolbar, choosing it from the View menu, pressing Alt+5 or double-clicking a MIDI Clip.

Although the Piano Roll is very flexible and configurable its 'normal' look is that of a keyboard down the left-hand side and a series of corresponding 'cells' to the right. If you've ever seen an old fairground organ that uses punched card to play its tunes then this may be a little more familiar in use. Each of the keys has a corresponding 'lane' extending to its right and these lanes can contain blocks of data that correspond to each key's pitch. Time is shown in the horizontal plane so a note that lasts one bar played on the C4 key will appear as a block one bar long in the C4 key's lane.

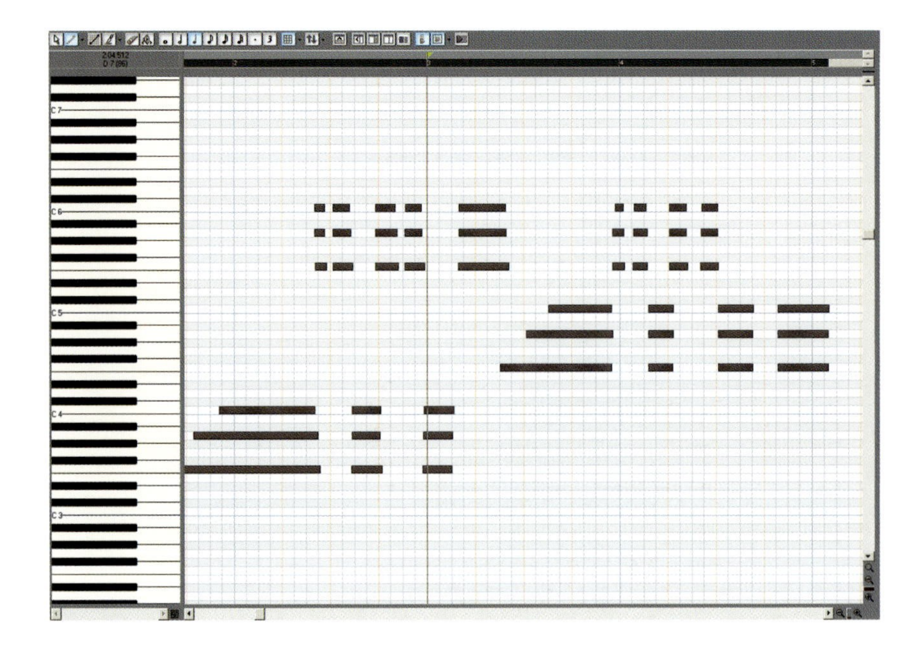

A set of tools is available in the upper left of the view with the Select ( ), Draw ( ) and Erase ( ) tools being the ones most used for note Input.

Note values are selected by highlighting a note and adding a dotted or triplet value if necessary from the selection to the right of the tools ( ).

To insert a note select the Draw tool, click on a note value, then simply click in the correct lane and position of the Piano Roll, if you have a sound source attached you'll hear the note sounded as you insert it.

If it doesn't go where you want it to then you can use the Snap To options to make it snap into place correctly or even turn Snap To off ( ).

Click the arrow next to the Snap To button to access the options. When you insert a note it will 'stick' at the resolution you have chosen making accurate note placement easy.

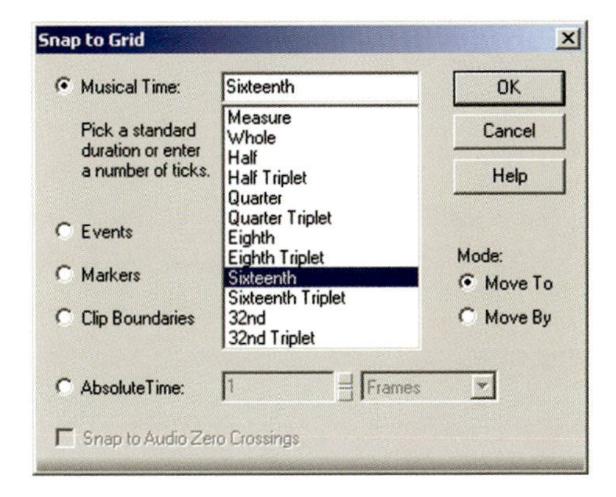

The Piano Roll will be covered in more detail in Chapter 6.

Inputting Notes in the Staff View

The Staff View uses normal notation methods to generate traditional style music scores. It can be invoked by clicking its button ( ) in the View Toolbar, choosing it from the View menu or pressing Alt+7.

Similar tools are available to the Piano Roll View and also note values can be selected in the same fashion but the Staff View also has options to fine-tune the note values by entering exact values in the Note Duration field (. 3 1:000).

Notes are entered using the Draw tool after selecting the relevant note value and simply clicking on the stave at the correct place. Rests will be automatically inserted as you go. Notes placed incorrectly can be dragged to a new position and their properties changed by right-clicking on the note head and changing parameters in the dialog that appears.

For guitarists a Fret View can be added by clicking this button ( ) from the available tools. When enabled you can position the song pointer at the correct place and insert notes by clicking on the fretboard. If you right-click on the fretboard you can even change the look of it to match your guitar neck or change it to a mirror image if you play left handed!

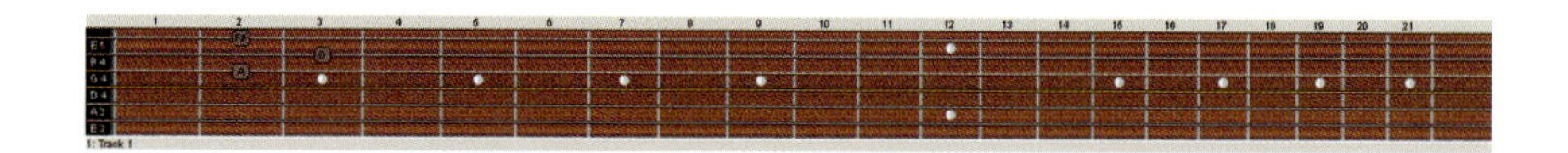

Recording Multiple Tracks

If your MIDI system is, or becomes, quite extensive you might want to record multiple MIDI Inputs simultaneously, which can also be particularly useful for recording several musicians at once who are using MIDI compatible instruments.

SONAR has the capability to do this quite easily.

All you need to do is to ensure that each MIDI source has its own unique transmission channel and each SONAR MIDI Track is set to receive on corresponding channels so if for instance you have a MIDI keyboard, MIDI guitar and MIDI drums you could have the keyboard on Channel 1, the guitar on Channel 2 and the drums on Channel 4. Providing the Tracks recording them are also on Channels 1, 2 and 4 you will record each instrument onto its own Track.

In order to hear each Channel you also need to make sure that all their Input Echo buttons are set to On ( ). The active Track will look like this but you'll have to click each button until it turns green like this ( ) when using more than one Track.

You can play and record several sounds at once using this method by adding MIDI Tracks and setting them up as different Channels and sounds as in this example . . .

1 Insert the Cakewalk TTS-1 DXi and enable it to be played from your keyboard as in 'How to get sound out with a soft synth (p. 56)'.

2 Insert another MIDI Track by right-clicking in the Tracks Pane and selecting Insert MIDI Track.

3 Set this Track to also receive MIDI Omni at its Input and set its Output to Cakewalk TTS-1.

4 Now set its Channel to 2 (2-Cakewalk TTS-1), Bank to the first one in the list (15488-Preset Normal 0) and Patch to Strings.

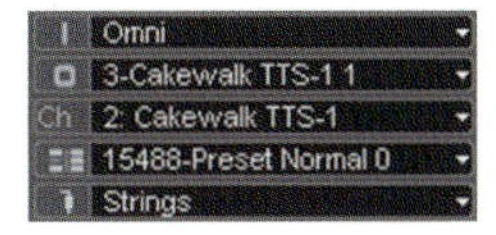

5 Click both Input Echo switches until they turn pale green (On).

6 Now when you play you'll hear the default Piano sound from the first Track and the Strings that you just set up playing together. They can also be recorded simultaneously by Arming both Tracks but can be edited or effected separately, maybe transposing one Track for instance.

Saving MIDI Files

MIDI songs can be saved in a variety of ways, the whole Project is best saved as a Normal Cakewalk Project file (.cwp) or as a Cakewalk Bundle (.cwb). If however you want to take the MIDI-only Tracks for use elsewhere (maybe on a floppy disk for use in a MIDI file player or keyboard) they can be saved as standard MIDI file types 0 or 1. MIDI Format 0 merges all Channels into a single Track while MIDI Format 1 keeps them separate. Both files are very small in size by today's standards. RIFF MIDI Format is also available for both types 0 and 1, RIFF stands for Resource Interchange File Format. Check the gear that you intend to use the file with for compatibility with any of these types but as the files are so small you should be able to try them out easily.

Working with Commercial MIDI Files

Many people use commercially available MIDI files to create their own backing tracks as most, if not all, of the instrumentation will be present in the file and depending on how well it has been produced can often result in an impressive sounding tune. Some sounds aren't easy to reproduce using MIDI such as guitars and vocals but you could always record these as audio alongside the MIDI Tracks if necessary.

MIDI files can be opened using Open a Project from the normal Quick Start dialog that appears when SONAR is started or by choosing Open from the File menu then browsing to the relevant file.

If you change the Files of type field to MIDI File then you will only see MIDI files in the Browser window making life a little easier.

Remember that the MIDI Tracks must then be routed into a sound module or soft synth, or they won't be heard!

I use the following method to get up and running quickly and then tweak things and add better sounding modules later:

1. Open your MIDI file in SONAR so that you can see all of the Tracks.

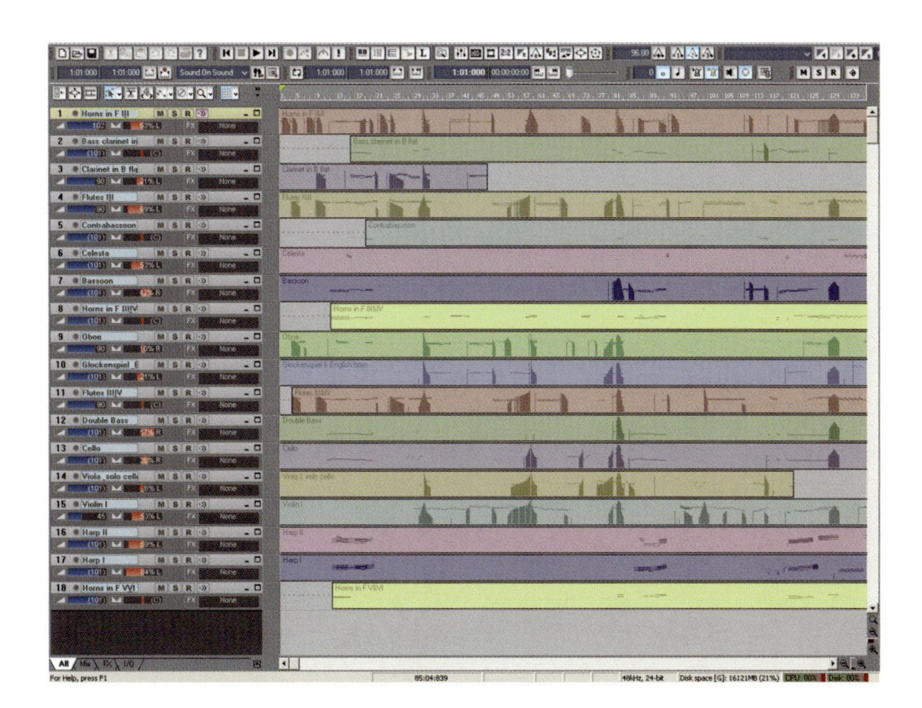

2 Use the Synth Rack to insert the Cakewalk TTS-1 but don't insert a MIDI Source Track from the Synth Options unless you intend to add more MIDI to the Project (☐ MIDI Source Track).

3 Press Ctrl+A or choose Select>All from the Edit menu.

4 Go to the Track menu and choose Property>Outputs.

5 In the MIDI Outputs field select Cakewalk TTS-1 from the drop-down menu. Click OK.

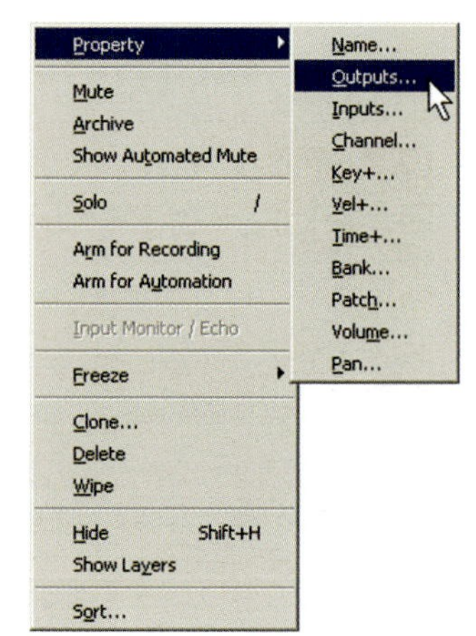

6 You can now play your file and all of its channels are routed through the TTS-1. If it has been configured properly then each Channel will automatically assign itself to the correct Patch when playback commences.

7 You can now change the Bank and Patch fields for each Track and also add other soft synths or modules and route each Track to specific places. The TTS-1 sounds can be edited by clicking the Edit button at the top of a strip to open the editor for that sound.

If you have a really nice soft synth, sampler or software drum module you could load it up then send any Tracks to it; it will also have its own audio Output (or even multiple Outputs if you load them from the Synth Rack) which enables you to add effects very specifically to just those Outputs.

Here's Native Instruments' Battery-handling drum duties from a standard MIDI file.

CHAPTER 5
EDITING AUDIO

Whatever you use SONAR for you will usually have to work with audio at some point. Even if you work with soft synths and MIDI for your Projects they will have to be converted to audio if you want to master them properly and burn them to CD within your computer.

In many ways audio may not seem as flexible as MIDI, it's not as easy to move a note for example, but audio does have its own set of advantages particularly with regard to using effects. In this chapter we're going to look at the variety of methods available for editing audio from simple cut and paste operations to more involved processes like Groove Clip Looping.

Recent advances in software technology have seen the introduction of non-destructive editing which is truly a revelation compared to any earlier methods. What it means is that you can make an edit such as removing noise from a quiet passage but if you later find out that you shouldn't have taken quite that much out as you've cut off a cymbal's decay for instance, you can simply restore the audio exactly as it was even after you've saved the Project and closed it. The audio is always still there just hidden and muted; providing you don't tell SONAR to apply the edit destructively, which is described below.

Editing in the Track View

The main Track View is the place that most of your audio surgery will happen which makes for one less View to have to worry about and also means that you can edit with relation to any other Track you like, MIDI or audio, by simply dragging them up or down from the Track Pane so that they are adjacent to each other. Just place the cursor over the Track's audio ( ) or MIDI ( ) identification icon so that it turns to a pair of vertical arrows ( ) then drag the whole Track up or down to your chosen destination.

Use of the Zoom tools is necessary for getting in close and I find pressing Z to invoke the Zoom magnifier cursor is a quick way of accessing a particular area without having to use the horizontal and vertical parameters separately. All you have to do is click and drag over the area you want to zoom in on and it will be magnified, you can repeat this to get in even closer.

There are other useful options available such as the Navigator View where you can resize and move the Viewing Rectangle to select the visible area in the Track View or you could press Shift+S to zoom in to the current selection. You'll soon find the best method to suit your way of working.

Another useful and easy to miss feature is the vertical scaling function which helps to identify very quiet sounds (such as background noise) in audio Tracks. This is achieved by dragging upwards in the vertical splitter bar of an audio Track which zooms in vertically on the audio in view. It's best to maximize the Track first

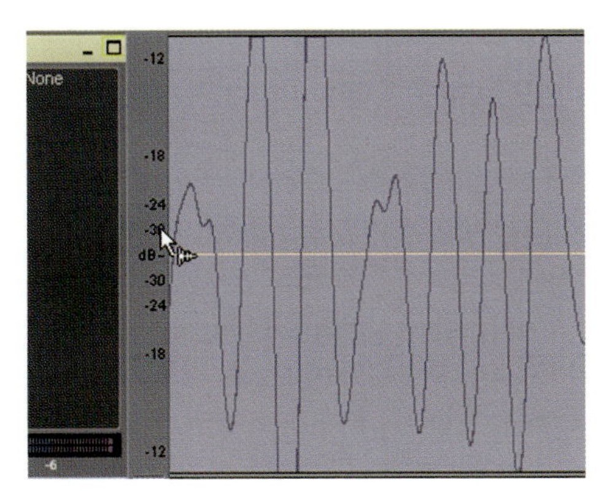

( ). Note that this operation doesn't affect the audio at all just the visual appearance of the waveform.

Cuts, Splits and Removals

The most common audio edit is to cut a section, usually to delete part of it or to copy the part for pasting elsewhere. It is very easy to do and can be achieved in a number of ways, the easiest being to use the Cut tool which looks like a pair of scissors and is invoked either by clicking on its button ( ) or pressing C. The tool is then simply placed over the audio to be cut and the mouse clicked, the edit will then be visible by its rounded lower corners. Note that the edit will be governed by your current Snap To setting so if you want to edit on a quarter beat you'll have to alter Snap To to a quarter or you can even turn it off to work right down to sample level ( ).

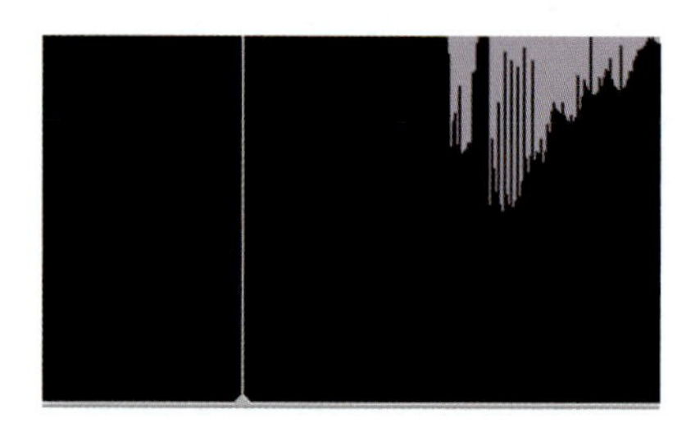

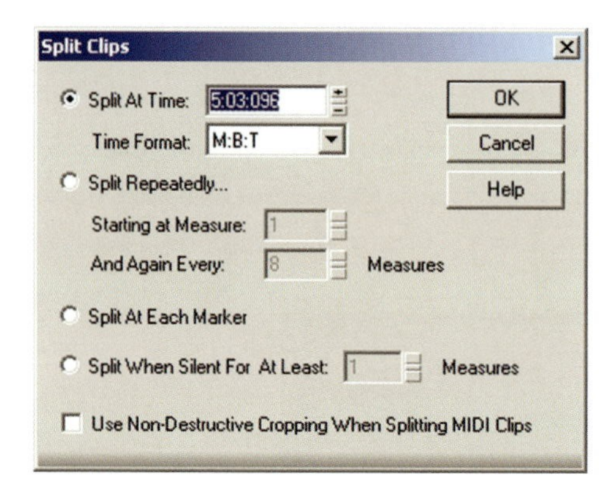

Another option for cutting is to right-click on the audio and select Split from the menu that appears, a dialog will open informing you of the options that you may want to apply to the edit. This is again governed by your Snap To setting as the cursor will obviously fall at a point compatible with the current Snap To setting but entering a different time in the Split at Time field will override it.

Deleting any audio section can be achieved by simply highlighting it and pressing the Del key. You don't have to split it out just click on the Track number to select the Track and then drag in the Time Ruler for the required length then delete the section.

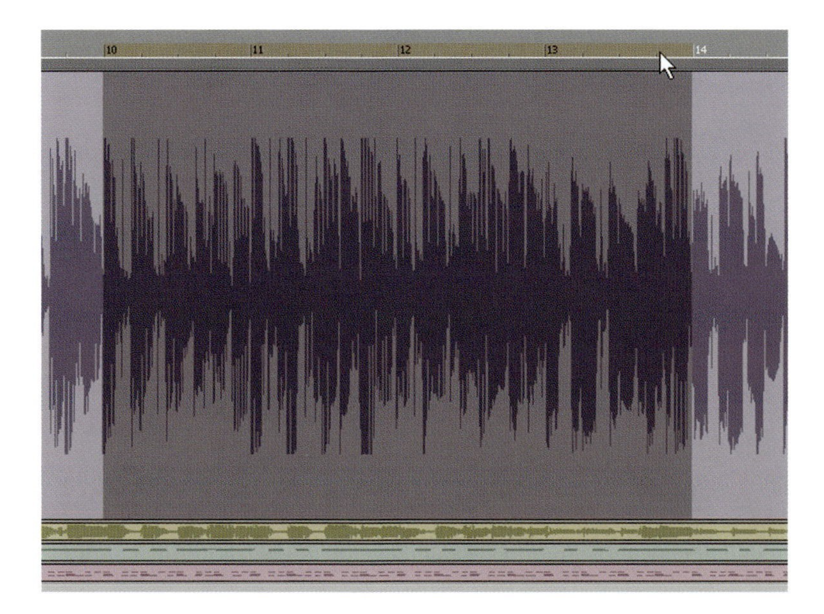

Holding down the Alt key while dragging on top of the Clip or Clips can also be used to make a selection.

If you want to be more selective about what's actually deleted as well as the audio (such as Automation or Tempo changes) then right-click on the selection in question and use the Delete command which will then offer you a dialog box with options about what exactly to delete and whether to delete the hole left by the edit. Note that only necessary options are available so if you don't have any Markers on the Track for instance the option will be grayed out.

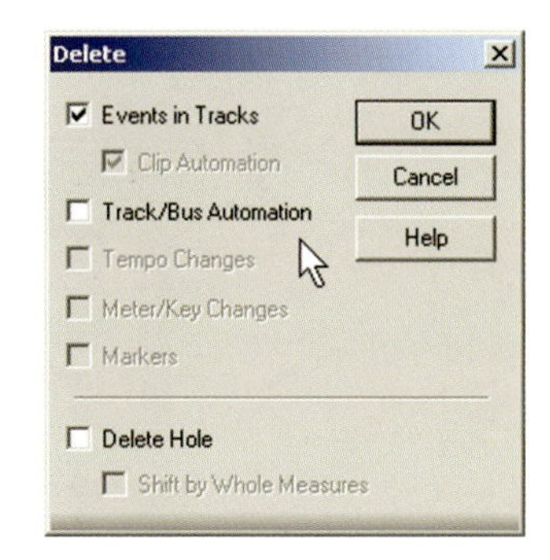

Non-destructive Slip Editing

If possible it's best to use non-destructive means to make edits so here's a little exercise to demonstrate how to do it and how it can save your bacon later!

1. Start or open a Project and make sure it has some audio content (you can Import a sample if you like).
2. Zoom in a little on the Clip and make a cut at any point.

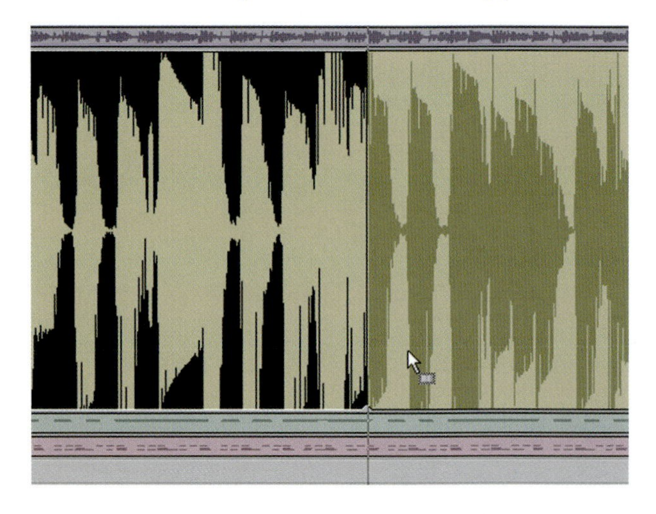

3. Place your cursor over the edge of this edit so that it looks like this. It will reverse depending on which side of the split it is.

4. Click and drag the edge of the Clip and you'll see that you can roll it out of the way!

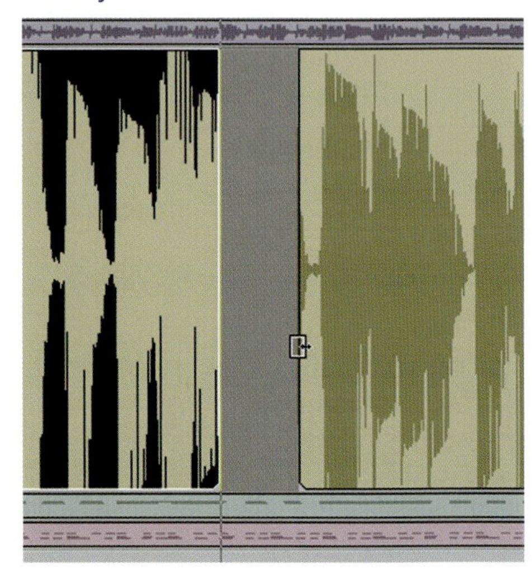

5 Roll some of the left Clip back and then highlight the right Clip and Delete it.

6 Oops! You shouldn't have deleted that Clip so you could use the Undo function to get it back but normally if you'd saved and closed the Project it wouldn't be possible to do that: If you've Slip Edited the Clip though you can now grab the right-hand edge of your edited Clip and drag it to the right.

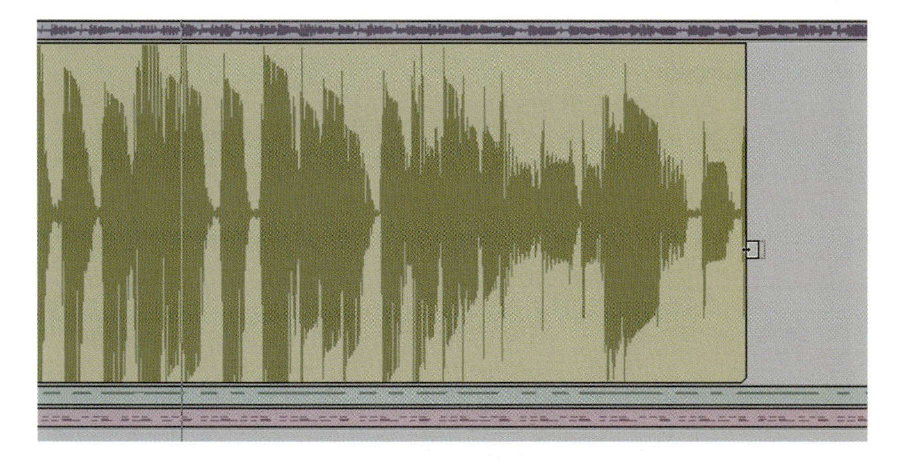

7 The deleted audio is still there and you could now redo the edit correctly if you wanted to, all this is possible even after saving and closing the Project.

Once you get used to this way of working there are an enormous amount of possibilities open to you which won't destroy any precious takes.

Fades and Crossfades

Another non-destructive and powerful feature is SONAR's 'draw-in' fades and crossfades, these are different to Automation which we'll look at later and are very quick to apply. I tend to use fades a lot just to ease the impact of audio kicking in and also to lock together the fade-outs at the end of songs when each musician has sustained the last note for a different length of time.

To Add a Fade . . .

1 Edit your audio to a suitable length by dragging its edge; the Zoom tools may be required to achieve accuracy.

2 Move your cursor to the top third of the Clip's edited edge so that it turns into this ( ) triangle shape for a Fade In or this ( ) for a Fade Out.

3 Click and hold the mouse button then drag into the Clip and you'll see the fade appearing.

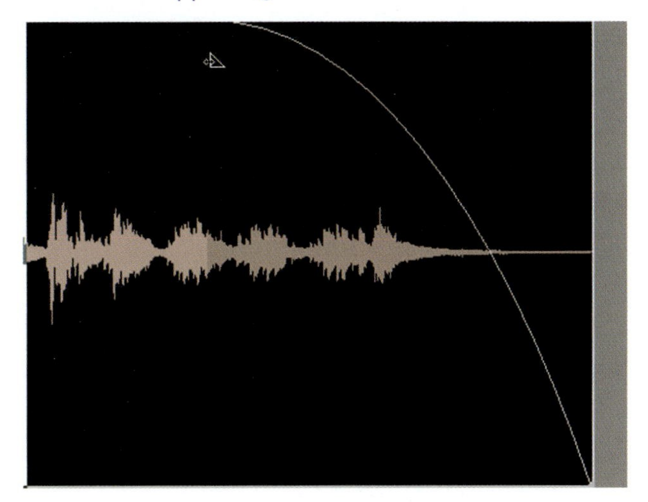

4 When you let go of the mouse button the fade will be set on top of the audio Clip.

5 You can make any fine adjustments to its length by 'grabbing' its edge (triangular cursor again) and dragging as appropriate.

6 With the cursor over the fade edge (triangular) right-click and you can now select a different fade shape if required to make the fade faster, slower or linear, try each one and listen to the different effect it has.

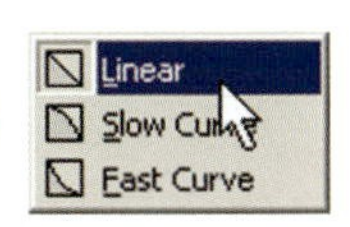

7 If you move the cursor to the lower third of the Clip's edge it can be used to roll the audio in or out while the fade stays applied and does not move with the edit. The cursor will look like this.

8 The middle third of the edge of a Clip can also be used to Slip Edit the Clip, including the Fade, but the Fade will move with the edit.

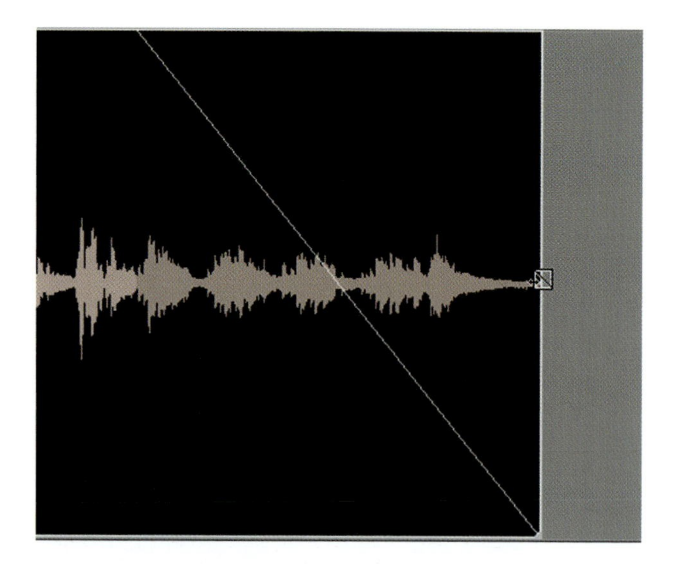

These edits can be performed at either end of a Clip.

You can also edit Slip Edit or add Fades to multiple Clips simultaneously by selecting them using Ctrl+click or Shift+click then dragging the edge of any Clip will also affect the other selected Clips. If you add or adjust a Fade then this will also affect all other selected Clips.

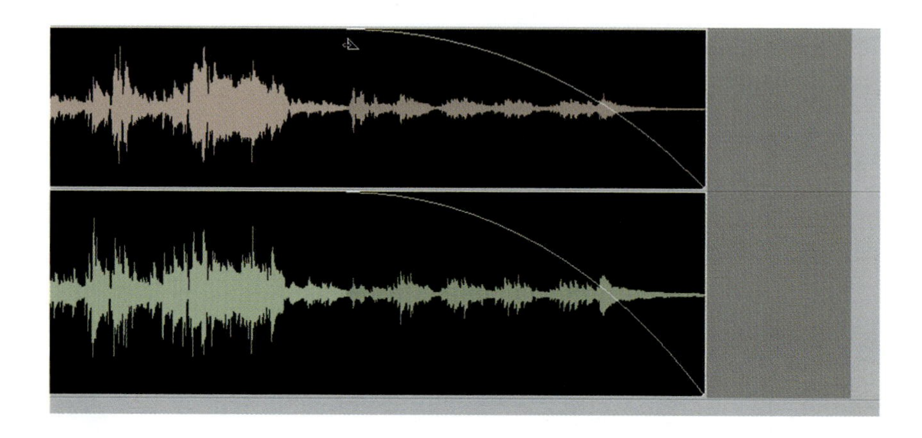

Holding down Alt+Shift will allow you to move the actual audio within the Clip!

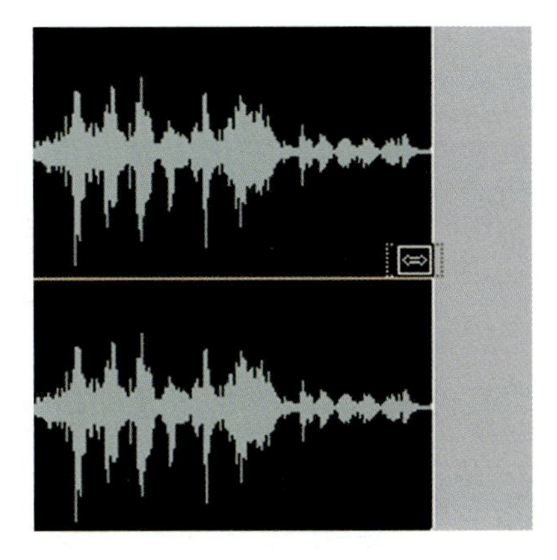

You can adjust the behavior of these edits by holding down the Shift key before performing the edit to change from an absolute edit to a relative one. Absolute will move all edits together while relative will keep them in their current relationships during the edit, that is a shorter Clip within the edit group will still remain shorter and move simultaneously with any longer Clips during the edit; using absolute, it would remain as it is unless the edit passed over it when it would be 'picked up' and edited accordingly.

For example, here are the Clips prior to editing:

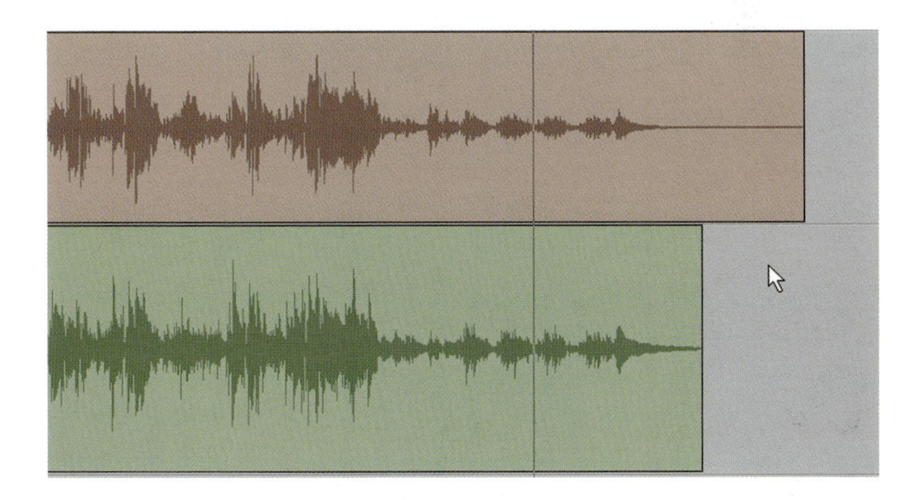

1 Absolute edit (default).

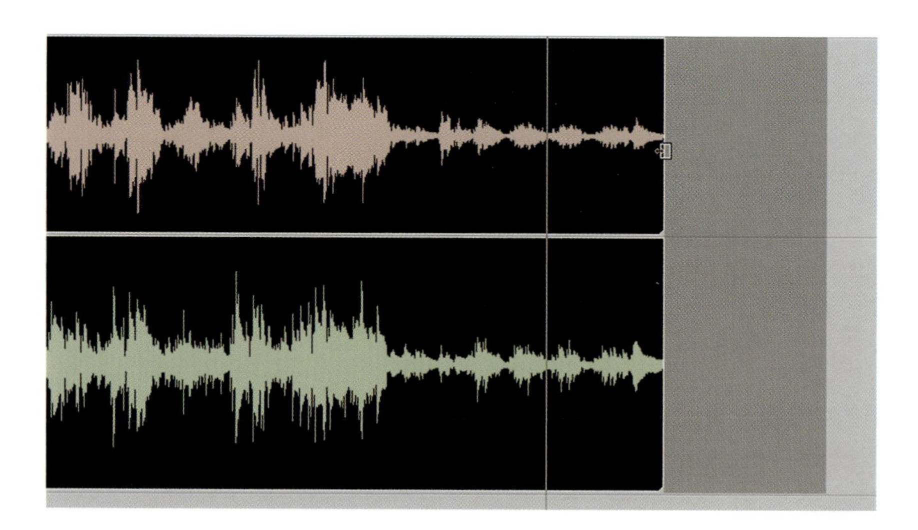

2 Relative edit (with Shift held).

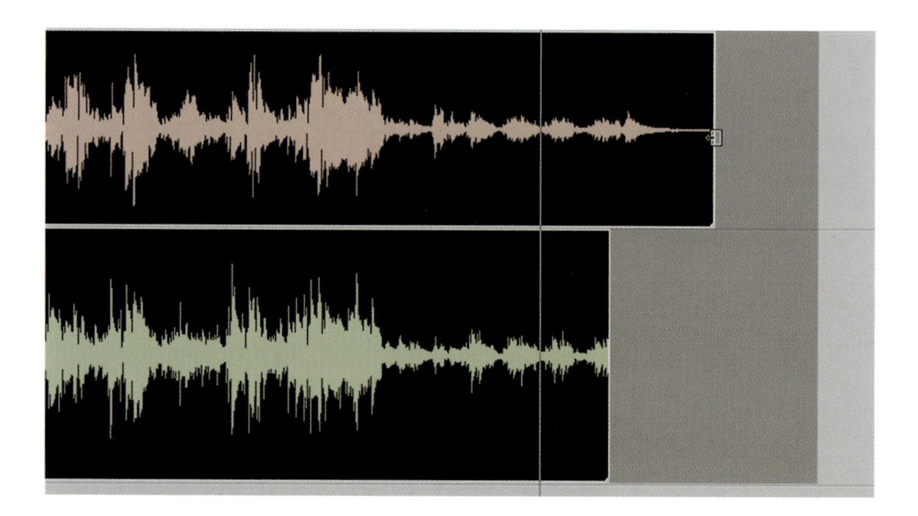

Crossfades can be automatically applied to overlapping audio Clips so that they sound seamless and blend together. This feature is activated by pressing X or clicking the Enable/Disable Automatic Crossfades button to the right of the Snap To button ( ).

The best way to try this out is to switch it on and then drag two pieces of audio so that they overlap on the same Track, the fade curves will then be visible and will also move to suit any movements of either Clip's edge during Slip Editing.

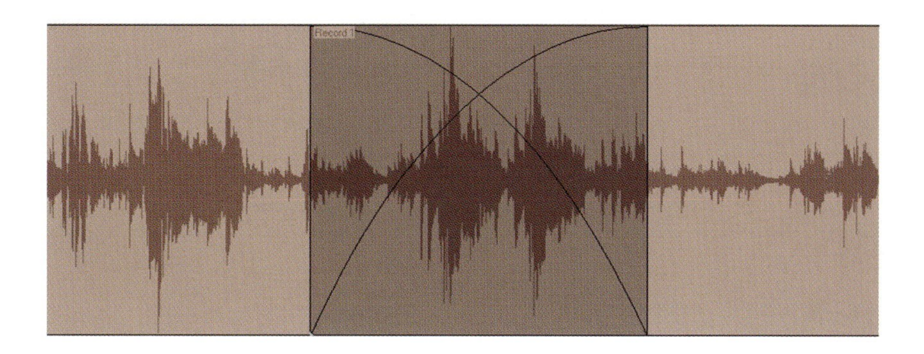

The arrow next to the Crossfade button provides a variety of default fade options for Fade-In Curve, Fade-Out Curve and Crossfade Curves, simply select one to make it your default for the current Project. Right-clicking on the crossfade envelope will also allow you to change its shape.

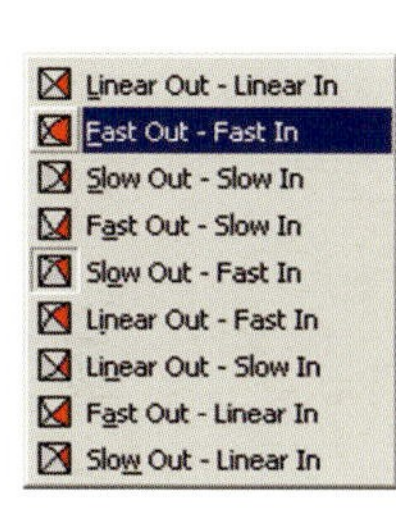

Other fades and crossfades are available from the Process menu but these require you to select the audio Track and define the length of time that the fade is required for first by dragging in the Time Ruler. The Process>Audio>Fade/Envelope dialog can then be invoked where curves can be customized by drawing in nodes and moving the shape around until it suits your requirements. They are applied as destructive edits and as such can't be undone once the Project has been closed so if you use them it's wise to make a copy of the audio Track and Archive it first just in case you want to change your mind but personally I'd use a non-destructive method whenever possible.

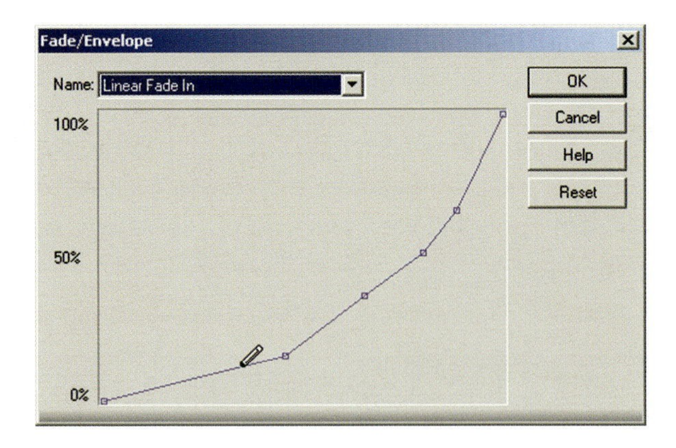

General Audio Features

SONAR has a range of features that can be accessed and applied destructively to a selected audio section, these features require you to highlight the audio section to be effected first and then they will be available from the Process>Audio menu or by right-clicking on the chosen audio.

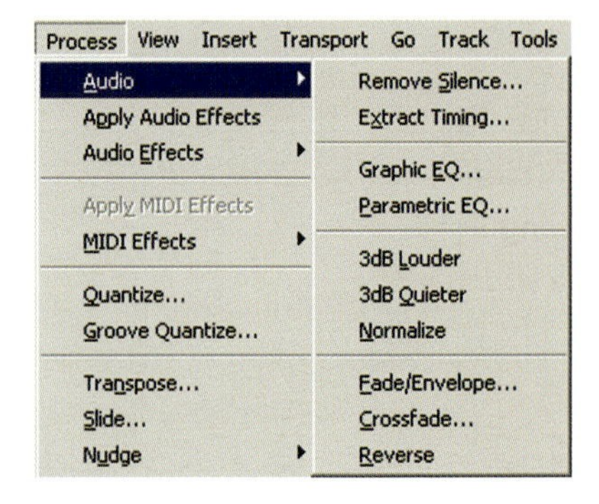

Also from the bottom of the right-click menu (or by pressing Alt+Enter) is the

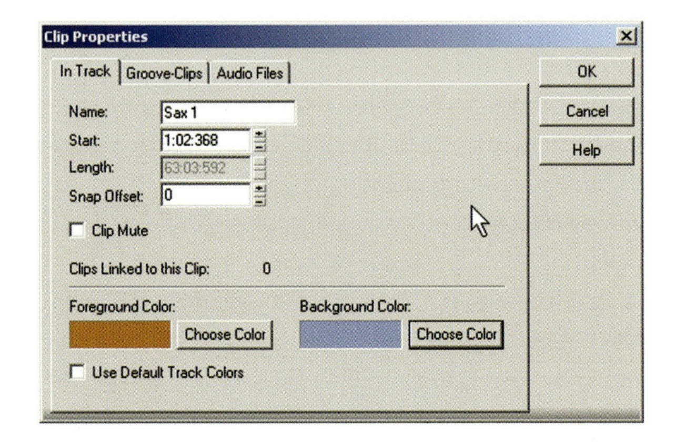

Clip Properties dialog which can be used to name the Clip, mute it and change its color among other things.

Effects can also be applied destructively from the Audio Effects submenu or either the Process or right-click menus but auditioning them is possible before application by clicking the effect's Audition button. If the audition time is too short you can adjust it in the Options>Global>General tab by increasing the Audition Commands for time in seconds (Audition Commands for 5 Seconds).

Increasing Level

Apart from the usual Volume and Trim controls SONAR has a variety of ways to increase (or decrease) audio level for a given Clip or selection. The Process menu has options to increase the selection a little by using 3-dB Louder or decrease it by the same amount using 3-dB Quieter. The option to Normalize is also here and this will increase the audio selection's level to its maximum without clipping, useful for quiet or badly recorded audio. There are also a range of dynamic level tools available using Automation which we'll look at in a later chapter.

Destructive Edits (Bounce to Clips)

Before applying a destructive edit consider whether you may need the original take again and if there's a possibility that you might, make a copy of it as follows . . .

1 Select the Track by clicking on its number (1), if you keep clicking it will toggle between selecting all, none and just the Track that you are clicking on.

2 From the Track menu select Clone and choose what you want to copy from its dialog. Click OK.

3 An exact copy of the Track will be made which you can now Archive (right-click on the Track's number and select Archive) and hide from view (same menu choose Hide Track).

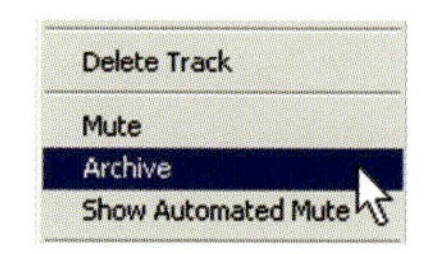

When you have performed a series of edits you might want to make them permanent and apply them destructively particularly if you are creating a sample for export or maybe trimming the ends of a completed stereo master. The quickest way to do this is to . . .

1 Make sure you've completed your edits and are happy with them.

2 Click on the Clip to select it.

3 Go to the Edit menu and select Bounce to Clips.

4 The audio will be rendered as a 'solid' edited Clip and any Slip Editing will now be destructively applied so the lower corners of the Clip will be square again. Any Fades drawn on the Clip will also be applied destructively.

Non-destructive Edits (Bounce to Tracks)

An option is available that allows you to Bounce any number of Tracks non-destructively to a new Track (or Tracks) which has many uses. If you have created a song that has many audio parts and is struggling to play you could mix and Bounce down some of them into stereo submixes then Archive the originals. You could create a stereo audio Track from soft synths without having to play and record them in real-time and you can also Bounce a whole song to a single stereo Track when it's finished. Here's how it's done . . .

1 Select the Track or Tracks that you want to Bounce down by clicking them. You can make multiple selections by holding Ctrl or Shift while clicking.

2 Make sure they are selected for the correct length of time by clicking and dragging in the Time Ruler.

3 From the Edit menu choose Bounce to Tracks.

4 A dialog box will appear where you must make mix selections. These are fairly straightforward as the Destination is the Track where your Bounced

Tracks will go (or start if there are several), Preset is any saved settings, Source Buses/Tracks are what you've selected, Source Category defines what you'll get, so Tracks will provide separate Track Bounces, Buses will provide any separate Bus mixes, Main Outputs will be separate mixes of multiple Mains if you have them and Mix will be everything selected Bounced down as a single file, the type being defined by the Channel Format field. Mix Enables provides options to further fine-tune the results.

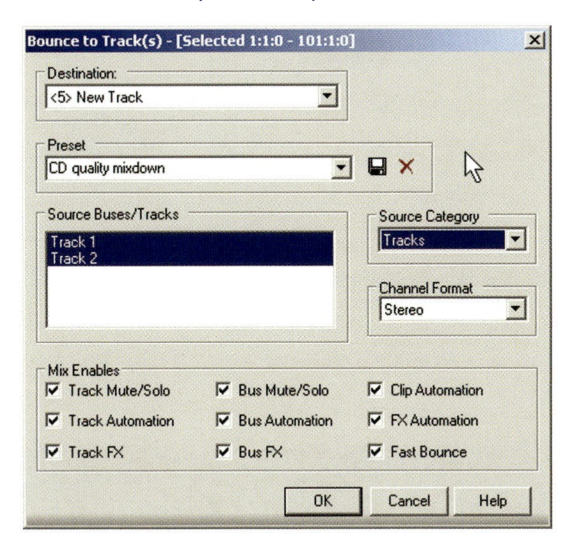

Comping Tools

Often when recording you may make multiple takes in order to get the very best from the performer; a vocalist may for instance sing slightly different on each take but overall you would have a very good take if you edited (or compiled) the best parts together. SONAR has a set of tools that can help you with this process by enabling you to selectively Mute any part of a Clip or even a whole Clip depending on your selection. We'll cover some of these topics in more detail in Chapter 12, Mixdown and Mastering.

The Mute Tool

This tool resides at the top of the Tracks Pane and looks like this ( ). It can be invoked by clicking its button or by pressing K and can be particularly effective and easy to use if you put all of your takes into either a Track Folder or make them into Track Layers on multiple lanes running on a single Track. Its menu provides options for Click+Drag Behavior which means the way the tool is implemented when dragged over a Clip.

The top option (which is the default) will Mute the Clip when you drag in its lower half; the tool will look like this ( ). When you drag in the upper half of the Clip the sound will be unmuted over that range; tool looks like this ( ).

✓ Mute Time Ranges (Alt+Drag Mutes Entire Clips)
Mute Entire Clips (Alt+Drag Mutes Time Ranges)

If you hold down the Alt key while dragging then you'll Mute the whole Clip (or unmute it if it's already Muted).

The lower option in the menu is the opposite of this and dragging on a Clip will Mute the entire Clip, while holding Alt will mean that you only Mute the range that you drag in.

When a selection is Muted it will appear as a dotted outline (audio) or grayed out (MIDI) so that it can be easily seen.

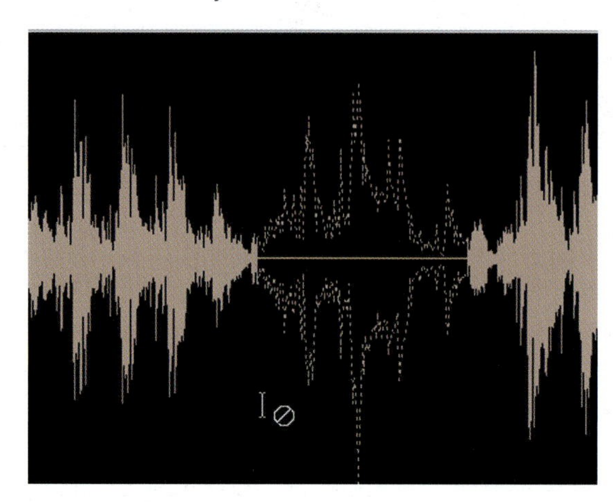

Selective Clip Auditioning

If you have each section recorded or edited as a separate Clip then you may want to audition specific Clips from different takes or Tracks, this can be achieved by holding the Ctrl key and selecting the Clips you want to hear then holding down the Shift key while pressing the Spacebar to start Playback. You'll only hear the Clips that you selected playing.

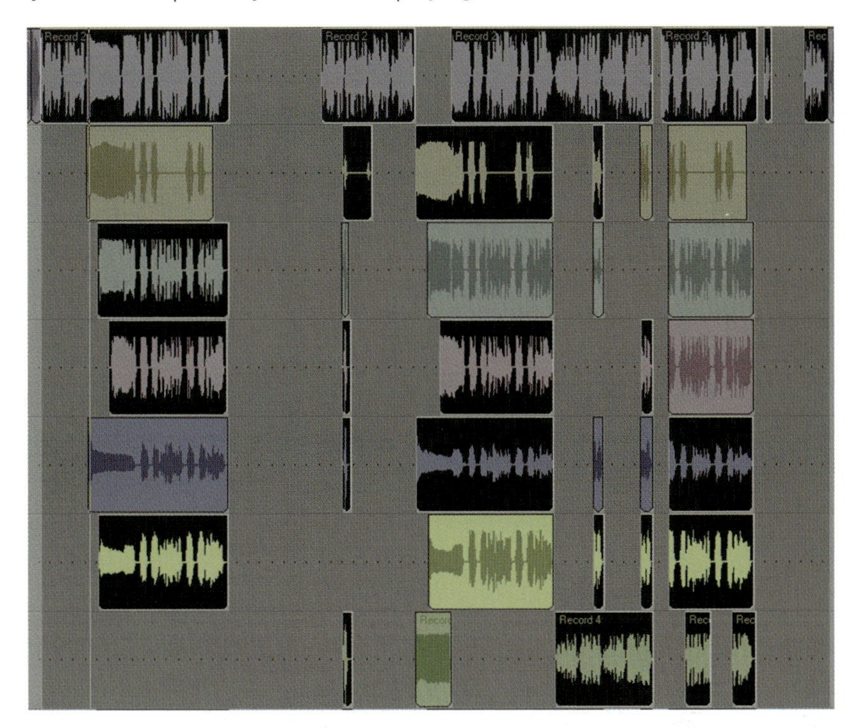

Creating Loops (Groove Clips)

A fantastic feature of SONAR is its implementation of Groove Clips also known as Acidized files as the technology was originally created by a company called Sonic Foundry for their sequencer ACID. What it basically allows you to do is create a piece of audio, a drum pattern for example, and then use it over many bars in your Project but there's a little more to it than that as Copy and Paste could achieve the same effect. Groove Clips can simply be dragged out or 'painted' into a Track and they can also follow tempo and pitch changes in a Project, which used to be the preserve of MIDI data.

The reason that they can follow tempo and pitch is because they have an identity which defines their default tempo and pitch and, based on this, calculations can be automatically made to compensate musically for these changes.

Drum and percussion Groove Clips don't usually have a pitch identity (although they can) so when a pitch change occurs they don't follow it, their pitch is however automatically adjusted for any tempo changes so that they don't sound higher when things speed up or lower if they slow down.

Musical Groove Clips will have both identifying parameters usually so that they respond correctly to match any variation of tempo or pitch.

Groove Clips can be Saved outside of SONAR but still retain their characteristics and as such can be Imported from external sources such as Loop Sample CDs. To start working with Groove Clips we're going to look at Importing them from a CD first using the special Loop Explorer View.

The Loop Explorer

To open the Loop Explorer you can click its button in the View Toolbar ( ) press Alt+1 or select it from the View menu.

The View itself is much like the standard Windows Explorer but is specifically aimed at finding and auditioning audio samples anywhere on your system whether on hard drive or removable media such as a sample CD.

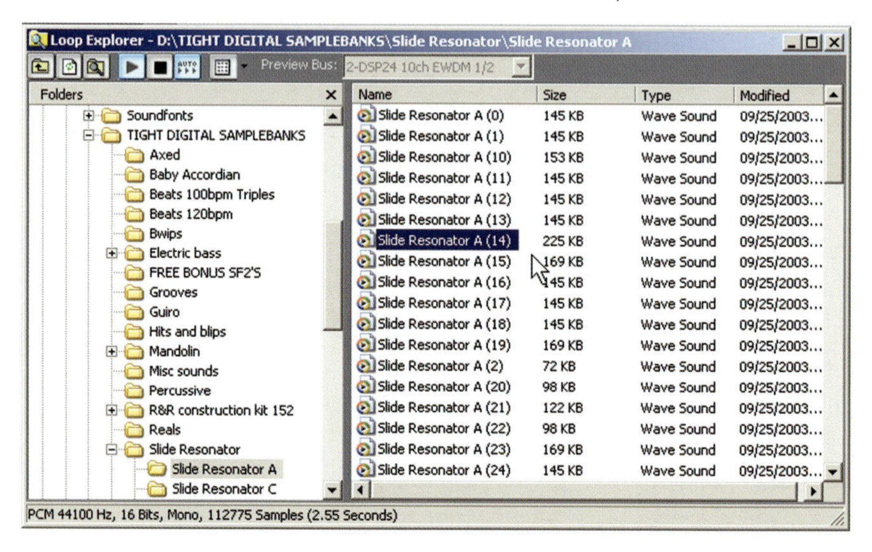

There are a few standard buttons for moving up one level, refreshing the view and choosing the view type but there are also a few extra ones which enable you to Play samples ( ). Stop them playing ( ) and make them Auto play ( ) when clicked.

The Preview Bus menu can be used to select which of your available Buses will be used to play the sound from during audition. If you have a stereo-only card then you may want to set this to your soundcard's main output.

Auditioning Samples in the Loop Explorer

1 With the Loop Explorer open, use the browser in the left-hand pane to find a folder containing samples. The SONAR install CD has a folder full of Audio Loops. Browse through to a folder you like then . . .

2 The samples will appear in the right-hand pane.

Name	Size	Type	Modified
147 FILTERY ELECTRO 1	566 KB	Wave Sound	05/13/2001..
147 FILTERY ELECTRO 2	566 KB	Wave Sound	05/13/2001..
147 FILTERY ELECTRO 3	565 KB	Wave Sound	05/13/2001..
147 FILTERY ELECTRO 4	565 KB	Wave Sound	05/13/2001..
147 FILTERY ELECTRO 5	565 KB	Wave Sound	05/13/2001..
148 ALL SUBS LOOP 1	561 KB	Wave Sound	05/13/2001..
148 ALL SUBS LOOP 2	561 KB	Wave Sound	05/13/2001..
148 ALL SUBS LOOP 3	561 KB	Wave Sound	05/13/2001..
148 ALL SUBS LOOP 4	561 KB	Wave Sound	05/13/2001..
148 ALL SUBS LOOP 5	562 KB	Wave Sound	05/13/2001..
149 ELECTRONIC FUNK 1	558 KB	Wave Sound	05/13/2001..
149 ELECTRONIC FUNK 2	557 KB	Wave Sound	05/13/2001..
149 ELECTRONIC FUNK 3	558 KB	Wave Sound	05/13/2001..
149 ELECTRONIC FUNK 4	557 KB	Wave Sound	05/13/2001..
149 ELECTRONIC FUNK 5	558 KB	Wave Sound	05/13/2001..

3 Click the Auto button so that it is down and make sure your Preview Bus menu is set to a suitable output.

4 Click on a sample and it will play!

5 As long as Auto is turned on you can keep clicking samples and they will play, you can also hold Ctrl and click more than one sample so that they both play together. If you are using Groove Clips they will keep playing until you hit Stop or deselect them.

6 Note that Groove Clips will also play at the current Project Tempo and Pitch. You can have a song playing in SONAR while your audition and the loops will follow the Tempo and Pitch of the song as it plays.

Loading Samples from the Loop Explorer

1 When you intend to load samples from the Loop Explorer into a Project it's a good idea to resize both the Track View and Loop

Explorer View so that they are both easily visible, I usually have the Track View along the top half and the Loop Explorer along the bottom half of the screen.

Using the Tile in Rows option from the Window menu is a quick way to lay the Views out horizontally.

2 Place the Song Position Pointer where you want to place the sample and highlight the Track that you want it to go to.

3 Audition your sounds until you find the one you want then simply double-click it to send it to your pre-selected position.

4 Alternatively you can drag and drop the sample from the Loop Explorer directly onto the Track View to any position you like.

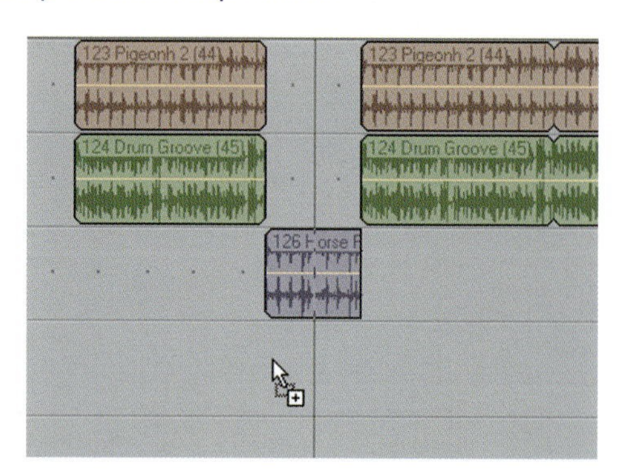

5 When Groove Clips are loaded into a Project you can then drag them out or 'paint' them to any length you want by grabbing the edge and dragging.

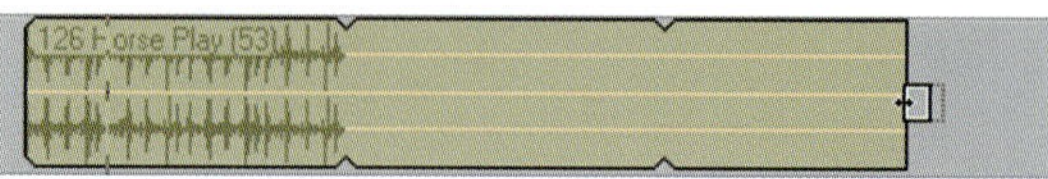

Making Your Own Groove Clips

In order to make a Groove Clip certain basic rules have to be followed, firstly the Clip has to be played in time with the current Tempo or it will not sound right when looped and you also need to know its Pitch if it is to follow musical Pitch changes.

Checking Your Loop

The best way to check if a piece will loop OK is to use the Loop Toolbar like this . . .

1. Record your audio to a metronome click or other timing guide, you don't have to edit it just yet as you may have several bars and wish to choose the best one.
2. Select the Track that your audio is On by clicking its number.
3. Click and drag in the time ruler for the duration you want the Loop to be, here's your chance to try out a few selections.

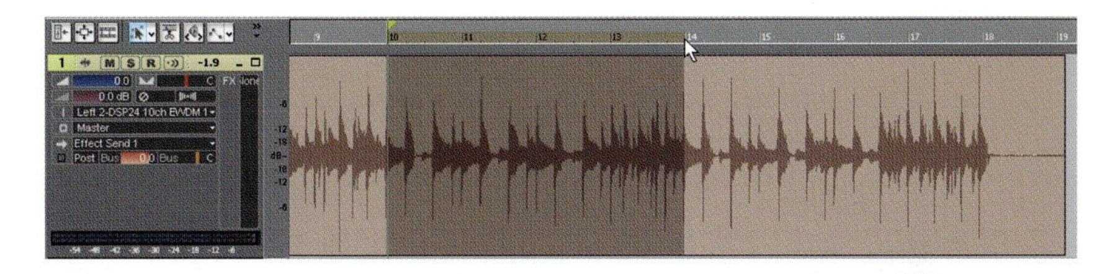

4. Click Set Loop to Selection in the Loop Toolbar (1:01:000 1:01:000), position your cursor at the start of it and hit Play, the selection will now play in Loop mode and you can hear if it works or not.
5. If you have a large 'take' to work from then you can try various sections of it to see which Loops the best.
6. When you have a good selection hold Ctrl then click on the selection (shaded) and drag it out to a free space where it will create a new Track. Use the Loop On/Off button in the Loop Toolbar to turn Loop mode off ( ).
7. You can now right-click on it and choose Bounce to Clips to make it a 'solid' sample but if necessary you should edit it first to make sure it plays smoothly as a loop. Make the loop fit into a logical time if possible such as four bars so that it loops easily on playback when used as a Groove Clip.

Converting a Sample to a Groove Clip

The quickest way to convert a sample to a Groove Clip is to click on it to select it then simply press Ctrl+L. If you haven't Bounced the Clip then you'll see a message warning that data outside the Clip will be lost, if you have still got the original or don't need it then click OK to proceed.

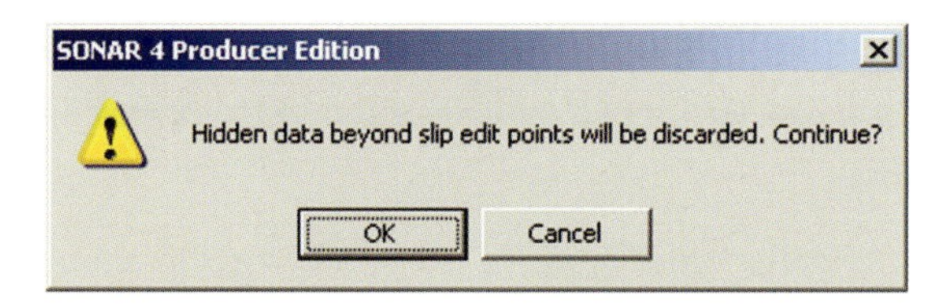

Most of the time this will work fine and you'll see all the corners of the Clip become rounded to show its Loop status. You can now roll it out for as many bars as you like.

Sometimes you may need to tweak the Looping parameters a little if the audio has some content that doesn't quite work and in this case we have the Loop Construction View to enable some fine-tuning of the audio sample.

The Loop Construction View

This View can be invoked using the Views Toolbar button ( ) pressing Alt+2, selecting from the View menu or by double-clicking on the audio sample. SONAR is usually very good at detecting audio parameters and will usually create Groove Clips correctly first time, if however it doesn't (often due to confusing transient peaks in the audio) you can help the detection using the Loop Construction View. You can also use it to twist and tweak your loops to give them a different tone or create other unusual effects.

The current Default Groove Clip Pitch is defined using the drop-down list at the right of the Markers Toolbar. You can change this to alter the Pitch of the whole Project.

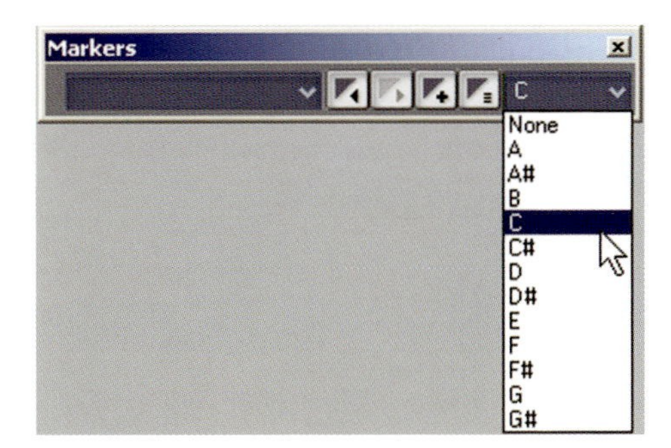

1. If your sample hasn't been converted to a Loop already then you can simply click the Enable Looping button ( ) or press L. If you get a dialog box warning you about hidden data beyond slip edit points being discarded it just means that you haven't bounced the audio down or edited it to size, if you're sure that you want to make the Loop permanent anyway then click OK.

2. When Looping is enabled you'll see the transient markers that SONAR uses to perform Loop calculations, click the Play button in the Loop Construction View and listen to the Loop.

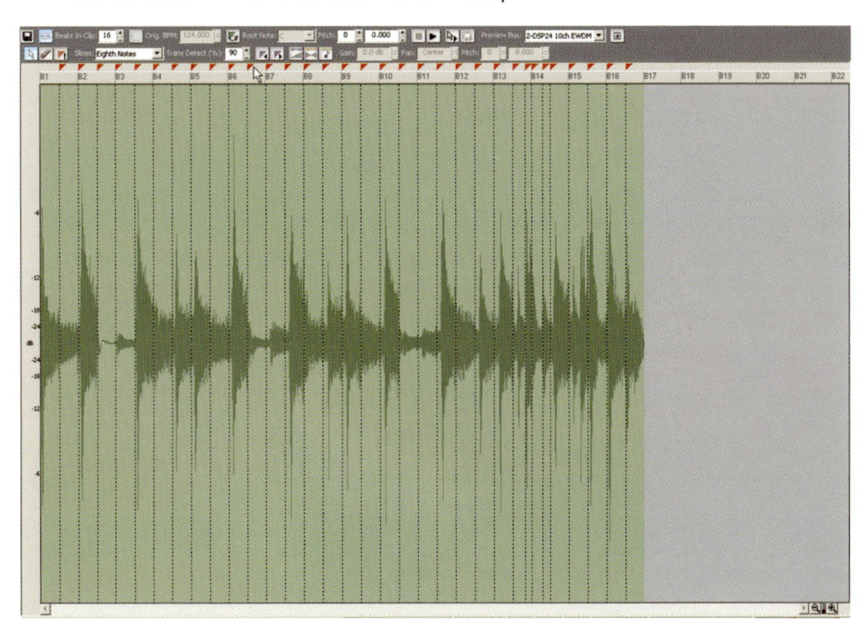

3 If the Loop isn't quite right then try adjusting the transient markers by dragging them or change the Slices parameter. The Trans Detect can also be adjusted to help SONAR automatically find the correct settings (Slices: Eighth Notes Trans Detect (%): 94).

4 Note that the Loop will play at the current Project Tempo setting and if you turn on the Follow Project Pitch button (Root Note: C) (or press P) then set the Loop's Pitch from the Root Note menu it will also follow the current Project Pitch.

5 The Beats in Clip and Pitch parameters can be used to correct problems but you can also use them creatively to change the speed and Pitch or tone of the Clip (Beats In Clip: 16).

6 Try turning on an envelope button for either Gain ( ), Pan ( ) or Pitch ( ) and then tweaking each Slice of the Loop by dragging the envelope up or down for some weird sounds!

7 To Save your Loop as a separate file just click the Loop Construction View's Save icon and you can put the Loop anywhere on your system for future use ( ).

8 If you want to keep working with the Loop within SONAR just close the Loop Construction View and carry on!

Easy Sample Creation

If you want to create audio samples (without the Groove Clip looping capability) from existing or new recordings then SONAR makes it easy as you don't have to carry out some of the 'usual' edits and it's possible to check the sample out for integrity before committing to a permanent edit. Here's how to do it . . .

1 First you need to start with a suitable audio Track, either mono or stereo will work.

2 Find the section that you want to take out as a sample and click on the Track to highlight it.

3 Drag in the Time Ruler (or press Alt while dragging on a Clip) to create a rough area containing the sample region.

4 Hold down the Ctrl key and then drag the shaded selection to an empty space where it will create a new Track. This copies the selection leaving the original intact.

5 From the Loop Toolbar click the Set Loop to Selection button.

6 Solo the new Track using the (S) button.

7 Now you can Play the Clip and see how it sounds, you'll hear how close you are to the correct edit as it loops around.

8 Zoom in on the sample so that it's easy to work on and then you can drag either edge in or out to make further edits, each time you do this click on the Clip again to select it then hit the Set Loop to Selection button again to update the looping points.

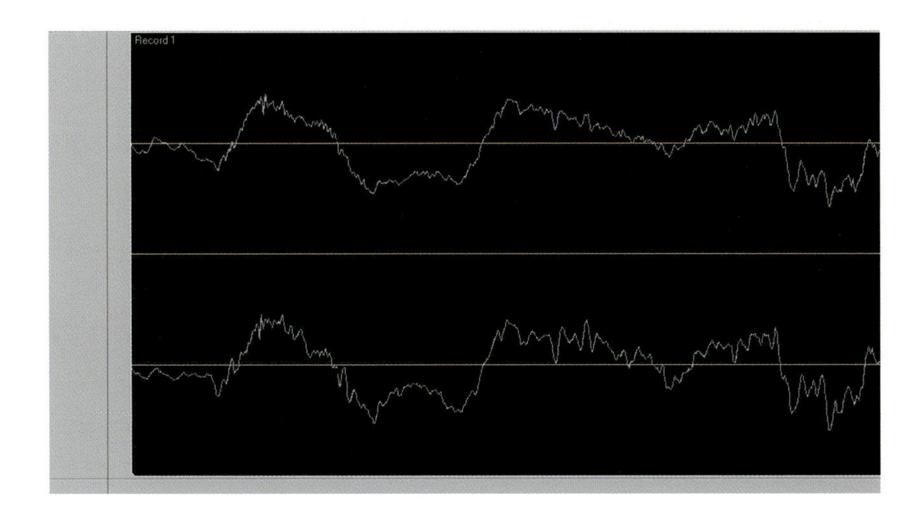

9 You may need to turn Snap To off to edit very closely. When you are sure the edit is correct you can also zoom right into either edge and add a small Fade curve to ease the audio start and finish levels a little.

10 When the edits are complete, highlight the Clip and select Bounce to Clips from the Edit menu.

11 If you want to Save the sample outside of SONAR you can use the File>Export> Audio dialog or simply drag the Clip onto your desktop where it will be automatically named after its parent Track.

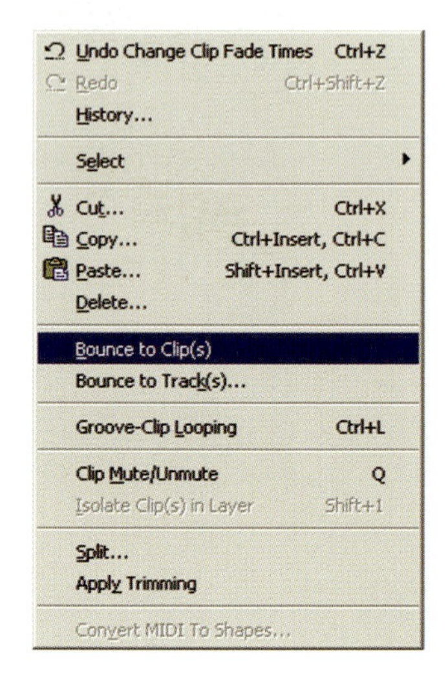

CHAPTER 6
EDITING MIDI

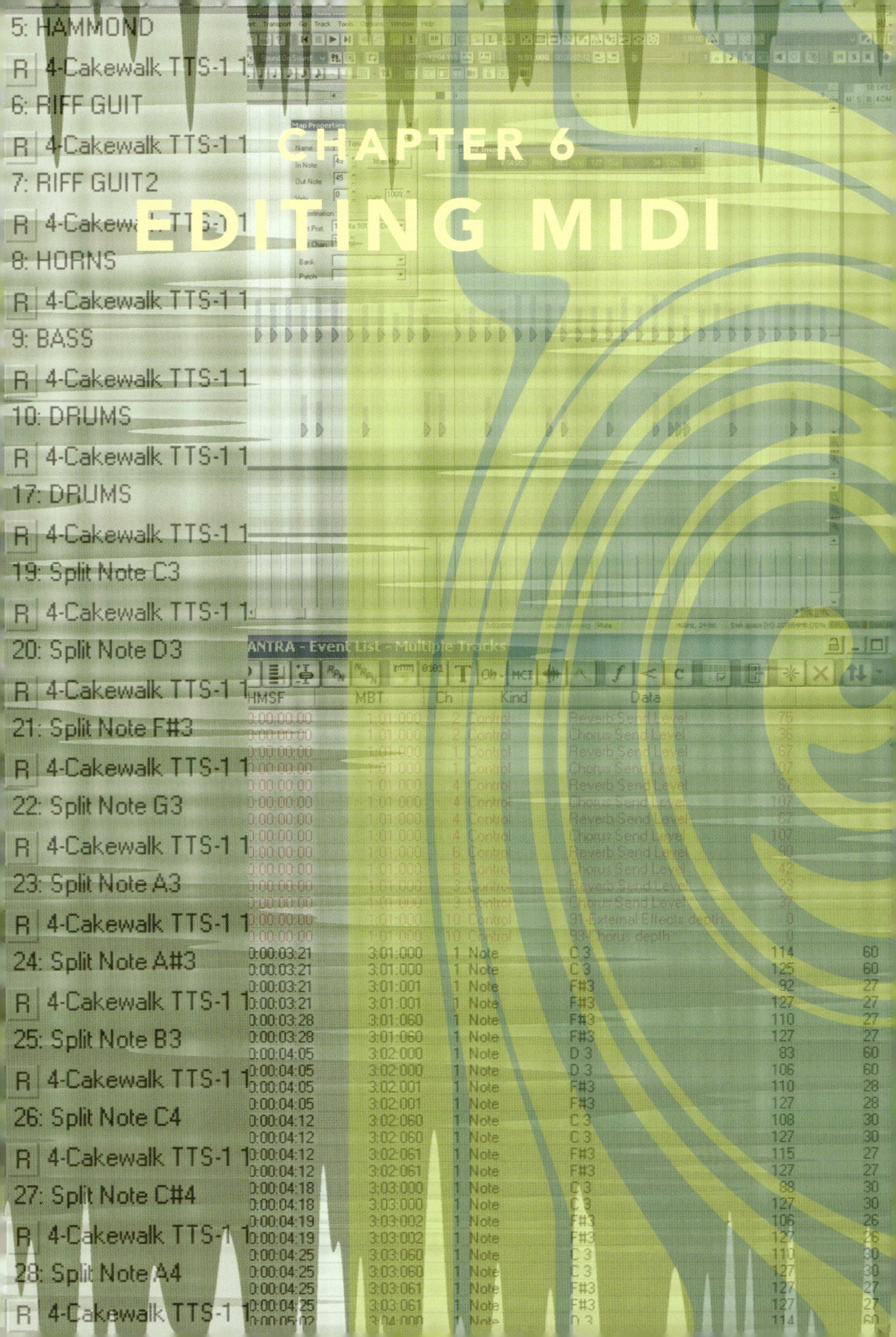

MIDI notes and data can be edited in a number of ways within SONAR using a variety of Views. Depending on your musical background you may prefer to use the Staff (notation) View or the default Piano Roll View as your main MIDI editor. There are also other Views, which can be used for more specific types of editing such as the Event List which is great for finding odd or spurious MIDI data, and the Sysx View for more advanced users to send and receive Sysx messages, useful for attached hardware. Another View that's not exactly a MIDI View but fits best in this chapter is the Lyrics View which can be used as an Autocue system after lyrics have been input, it can be particularly effective when a backing track is being played from SONAR as the lyrics will be highlighted in sync with the music.

MIDI Tracks can be saved as either Format 1 or Format 0 (the most common) or as RIFF MIDI Formats 1 and 0 by selecting the appropriate option in the Save As Type field when saving.

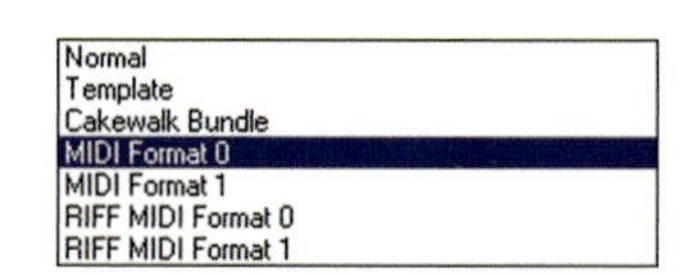

I'm going to concentrate mainly on the Piano Roll View as it is probably the most flexible one for working on MIDI in general and can be used to edit controllers easily. It also has a very powerful Drum Editing mode with specialized tools that make creating drum tracks easy.

If you double-click on a MIDI Clip it will open the Piano Roll View by default but if you prefer another View it is possible to change the default. Right-click on the Clips Pane and choose View Options to be greeted by the Clip View Options dialog. Simply select another option from the MIDI Clips menu and SONAR will remember the setting when you double-click on a MIDI Clip in future.

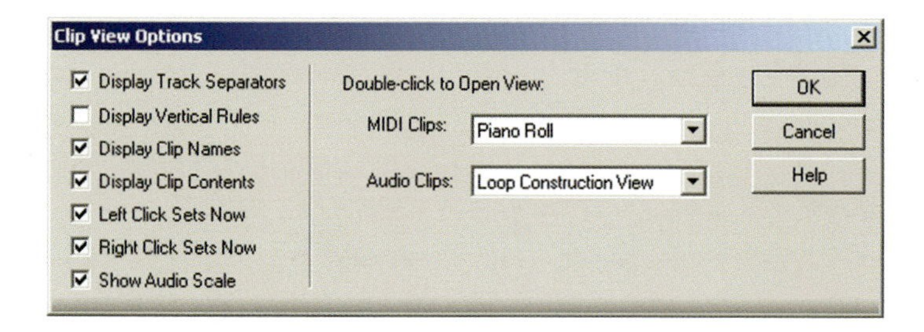

MIDI Groove Clips

In the same way as audio it's possible to create MIDI Groove Clips within the Track View by selecting the Clip and pressing Ctrl+L. The Clip will then show the same rounded edges as audio Groove Clips and can be dragged out along the Track to repeat as many times as you like. To Save the Clip outside of SONAR in this format just use the dialog from File>Export>MIDI Groove Clip. You can also right-click on the Clip and select Clip Properties then the Groove-Clips tab to enable looping and set its Groove parameters. These Clips will respond to Pitch Markers in the same way as audio Groove Clips.

Edits can be made to a single repetition of a MIDI Groove Clip without affecting the rest of the repetitions but if you roll back over it using Slip Editing then the edit will be lost.

In the Track Pane

Apart from the usual controls MIDI Tracks also have the following:

1. MIDI Tracks can be transposed by clicking on the (2) field and adding an offset (either + or −) to transpose in semitones.
2. MIDI Tracks can also have a time offset applied by clicking in the (0) field and adding an offset value.

If the Track's Output is going to a device that responds to common MIDI controller messages then you can adjust the Chorus and Reverb levels by dragging in their respective fields (Chr 45 Rev 77).

Editing in the Piano Roll

The Piano Roll View in its default format has a keyboard running vertically down its left-hand side while a grid to the right resembles the old punched card that was used to play fairground organs many years ago, long before my time! A similar principle is used however as blocks in the Piano Roll represent what used to be slots or holes in the card, corresponding to notes that are played as the song progresses. Time runs laterally and so any blocks that are short represent short notes while long blocks are longer notes. The scale along the top edge is the same as the Measures, Beats and Ticks that are the default in the Track View.

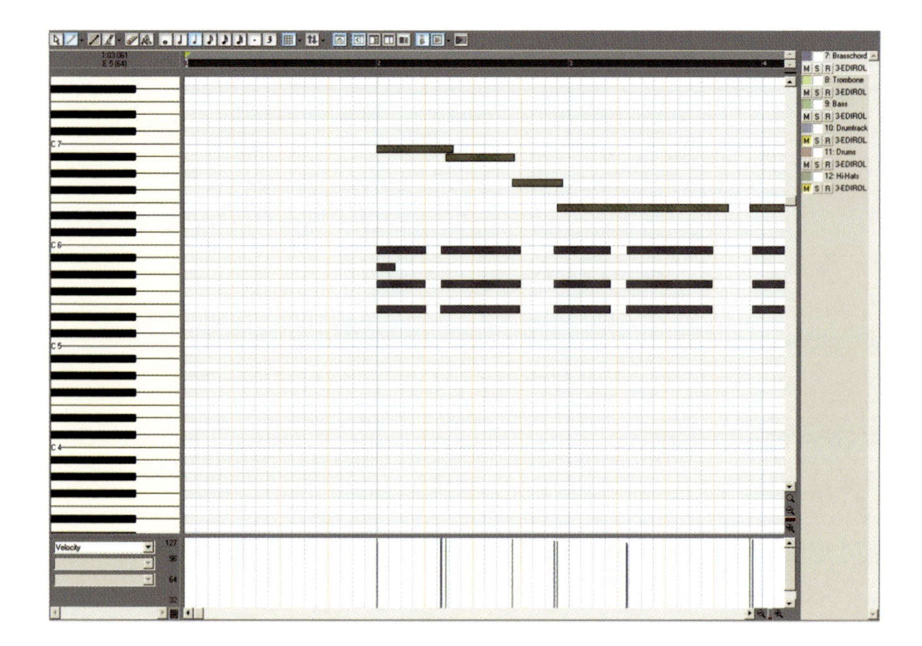

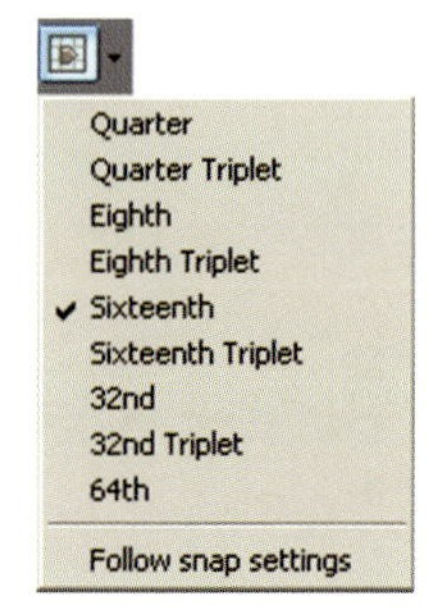

The grid is a useful guide to note lengths and can be set at a variety of resolutions by activating the Show/Hide Grid button and choosing a new setting from its drop-down menu.

A Snap To button is also available to help with placing and editing notes ( ).

Drawing Notes and Editing Them

Open the Piano Roll by double-clicking on a MIDI Clip or highlighting a MIDI Clip or Track then selecting the Piano Roll View from the Views Toolbar or menu ( ). The shortcut is Alt+5.

A selection of tools are available in the Piano Roll View, the most commonly used are the Draw tool ( ) [D], the Eraser ( ) [E] and the Select tool ( ) [S].

The Draw Line tool ( ) [L] is used for drawing shapes such as fades for controllers; you can also achieve this by holding down the Shift key when using the normal Draw tool.

The Scrub tool ( ) [B] can be dragged over the notes to hear them play.

You can select all notes of the same pitch by clicking on the relevant piano key or clicking and dragging across several keys, or

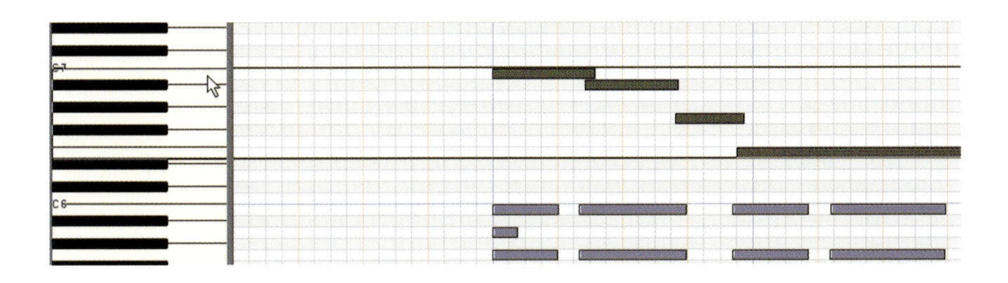

make a selection of a group of notes by dragging a lasso around them using the Select tool.

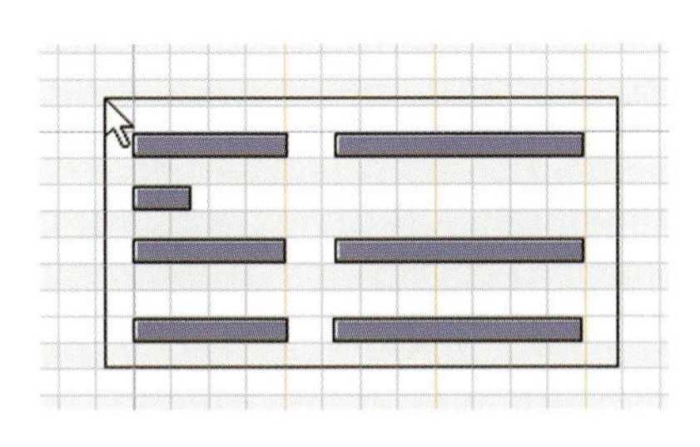

Note value can be selected using these buttons ( ); notice that you can also add a dotted or triplet value to any note:

1. To insert a note, select the Draw tool then click on a note value. Note values can be quickly changed using your number keys.

2. Click in the relevant cell of the Piano Roll and the note will appear and sound as it is inserted (providing the MIDI Track has been routed to an output source).

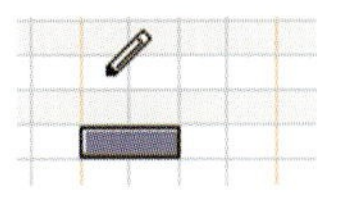

3. To move the note up or down, move your cursor to the middle of the note until it looks like this ( ) then you can click and drag up or down, and hear the different notes as you do so. This will constrain the

movement to the vertical plane so you can't accidentally move the note in time. Just release the mouse button to leave the note in its new position. You can also use the Select tool to move notes, holding down Shift will restrict movement to the horizontal plane.

4 To move the note in time (left or right) move your cursor just to the left of center so that it looks like this ( ) then you can click and drag the note left or right, this will constrain the movement to the horizontal plane so that you can't accidentally change its pitch.

5 To adjust the note length move your cursor to whichever end you want to alter until it looks like this ( ) then just click and drag to the desired length. The movement will be constrained to adjust length only so you can't accidentally move the note any other way.

6 Don't forget that you can use the Snap To options to help place your notes.

You can right-click on any note to check its Properties and change them if necessary.

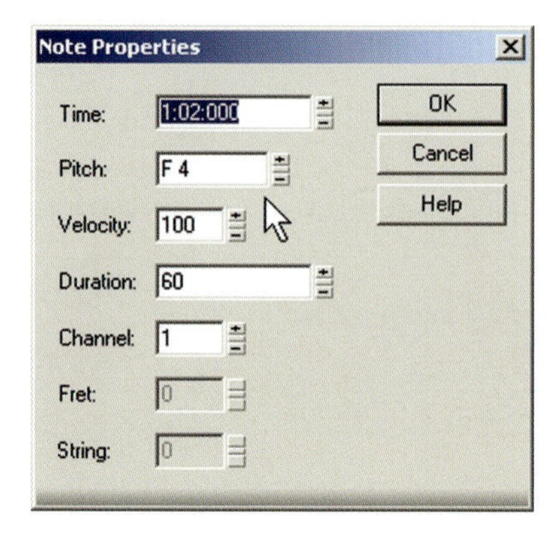

Drawing and Editing Controllers

As well as the notes themselves, at some point you will probably want to edit at least some of the other associated MIDI data such as Velocity, Volume, Pan or some of the more exotic controllers such as Filter Cutoff and Resonance. SONAR makes this an easy task by using an extension to the Piano Roll that is accessed by clicking the ( ) button just below the keyboard or ( ) in the strip of Toolbars at the top. This opens an expandable Controllers Pane with a set of

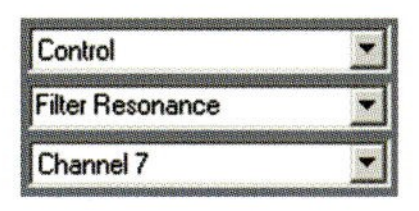

menus that are used to access specific controllers; the controllers available will depend on the Track's output so a piano soft synth will probably have different options to a polyphonic synthesizer. The top menu selects the controller type, the middle one is the specific controller (dependent on the type) and the lower one is the Channel currently in focus.

To make a simple Volume adjustment:

1. Place the Song Position Pointer (the Now Time) at the relevant place in the song, highlight the MIDI Track to be edited then open the Piano Roll from the Views Toolbar, use Alt+5 or double-click the Clip.
2. Open the Controllers Pane by clicking its button, pressing C or dragging the horizontal splitter bar upwards.
3. Choose Control from the upper menu, Volume (Controller 7) from the middle menu and ensure the lower menu is set to the correct Channel, that is, the one you're working on.

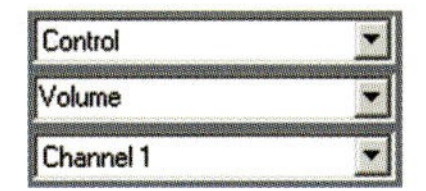

4. Select the Draw tool and click or click and drag in the Controllers Pane to draw. Notice that the value is visible on a readout at the top left of the view showing both time and current Controller value at the cursor point.

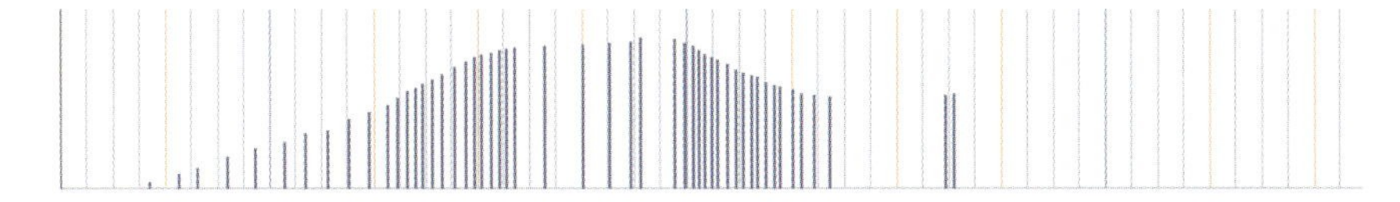

5. You can edit any part of your Controller envelope drawing using the Draw and Eraser tools.
6. To draw straight lines select the Draw Line tool click and hold the left mouse button then drag to the desired location and release. This is ideal for fade type envelopes.

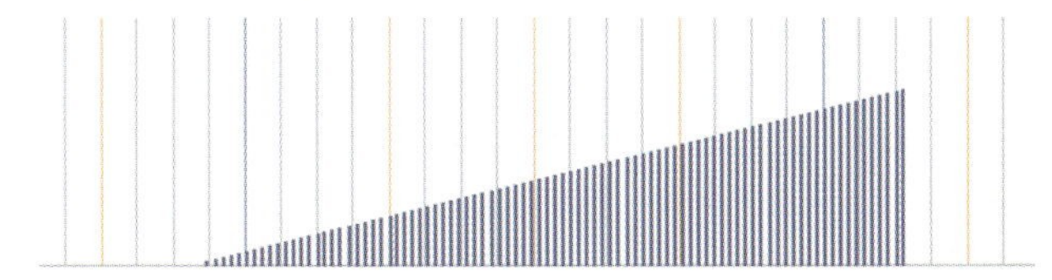

7 Be careful to draw in the correct position with regard to notes, a message will remain continuous unless you change it so if you draw a Volume fade down to zero then any following notes will not sound unless you draw in another value for them!

Multiple Track Editing

You may often benefit from seeing and editing several MIDI Tracks simultaneously, for example, violins and cellos. In this case you should use SONAR's Multiple Track Editing features which allow you to view and edit more than one Track but also have options to make this 'safe' to prevent inadvertent editing of data that needs to be seen but not edited. This all takes place in the Piano Roll View.

Using Multiple MIDI Track Editing

There are several ways to open Multiple Tracks in the Piano Roll View:

1. From the Track View, use Ctrl-click or Shift-click to select several MIDI Tracks then click the Piano Roll View button from the Views Toolbar or press Alt+5.
2. Open the Piano Roll View then use the Pick Tracks selector ( ) [T] to choose which Tracks to view. To select more than one use Ctrl-click or Shift-click. Its drop-down menu can be used to select either Previous or Next Tracks.
3. If the Piano Roll View is already open, Tracks may be added to it by using the Pick Tracks selector and using Ctrl-click to add to the existing selection.
4. You can also use the All Tracks ( ) [A], No Tracks ( ) [K] and Invert Tracks ( ) [V] buttons to change selections.

When you have several Tracks in view you can keep control over what you are editing and what is only visible (but uneditable) by using the Track Pane at the right of the View. This can be opened and closed using its button ( ) or shortcut key [H].

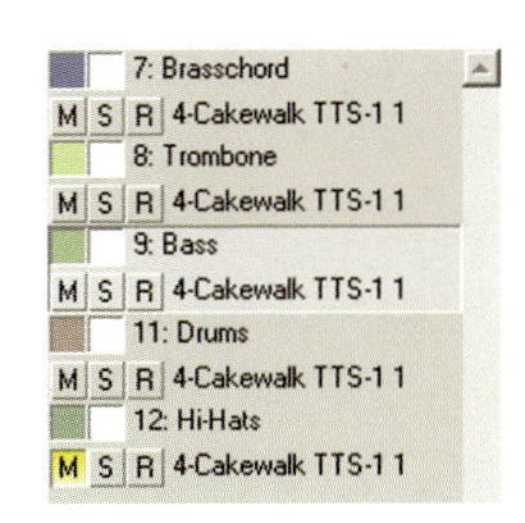

Each available Track is named here along with Mute, Solo and Record Arm buttons.

The currently selected Track is shown as a depressed panel, this will mean that any notes now added will be added to this Track so if it's a harmonica Track then you'll be adding harmonica notes.

To change focus just click on the name of another Track.

The colored square box at the left of the Track name can be clicked to Show or Hide the Track from view ( ).

The white square box can be clicked to Enable or Disable Track Editing. When Disabled the Track will appear 'grayed out' ( ).

Handy Piano Roll option – Although the keyboard at the left of the Piano Roll is useful you may find that other options, that is, names are better suited when working with some MIDI devices. To view and implement other names . . .

1 Right-click on the keys and a Note Names dialog will appear.

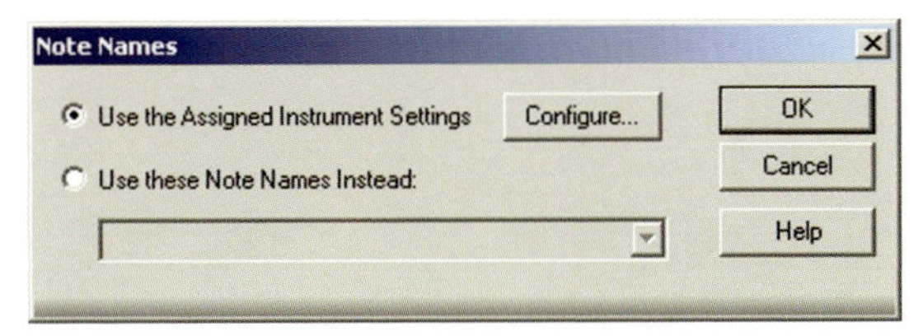

2 Check the box marked Use these Note Names Instead.

3 Make a selection from the list available in the drop-down menu.

4 Click OK and the names will replace the piano keys.

You can bring the piano keys back by checking the box marked Use the Assigned Instrument Settings and clicking OK.

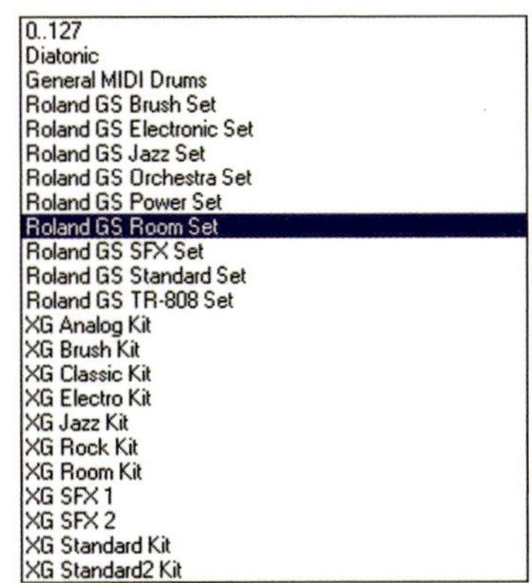

Many more options are available by clicking the Configure button in the Note Names dialog where Channels and Instruments can be Assigned, Defined and new Instrument Definitions can be Imported if required.

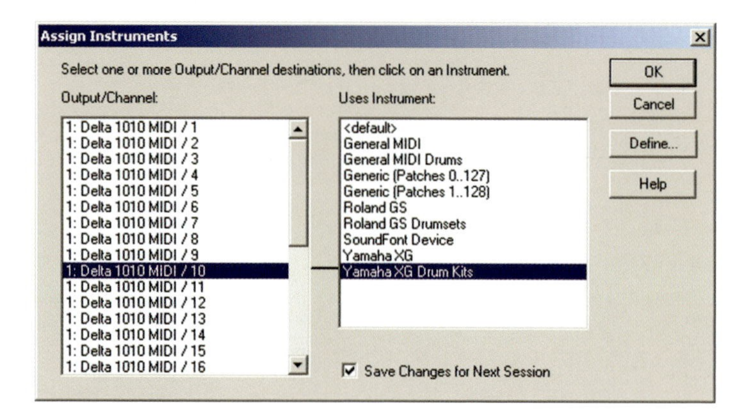

Editing in the Staff View

The Staff or notation View is a more 'traditional' version of note editing, consisting of a standard musical stave and notes that will be familiar to most musicians. It follows the accepted notation standards but also adds some further enhancements such as a guitar fretboard and the ability to Export Tab in ASCII format.

The main tools are much the same as in the Piano Roll View but the Staff View doesn't have a Controllers Pane so the Draw Line tool isn't present as it's not required.

A Pick Tracks selector enables multi-track editing to be carried out in much the same way as the Piano Roll.

The View can be magnified using the Zoom tools and it also has a Snap To facility to aid note positioning ( ).

Note resolution can be set using the selection available and fine-tuning is also available via the Note Duration controls (1:000).

Some note parameters may change depending on the status of the Snap To, Fill Durations ( ) and Trim Durations ( ) buttons as these will automatically override the input when activated. This is to help notate real-time playing without all of the minute details that are produced from a live performance; for instance a musician will play a quarter note closely but not exactly as a quarter note duration, this would appear unnecessarily complex as notation so the Fill and Trim tools can be used to simplify the appearance.

There are a wide range of options available for the Staff View from the Layout dialog available from this button ( ).

The Clef used can be changed to suit the instrumentation and there also exists a set of percussion definitions which are available when Percussion or Percussion Line Clefs are selected. These enable a variety of note bindings and shapes to be defined to suit the user.

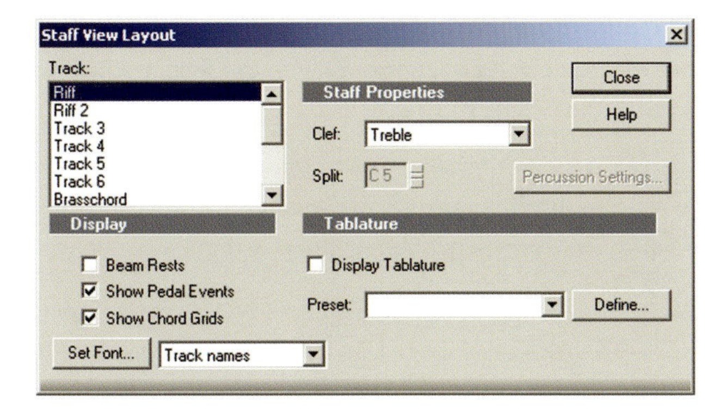

The Fret View can be added at the lower section of the View by clicking its button ( ). A right-click on the Frets will reveal several options including Mirror Fretboard for left-hand players and even a range of finishes to match your own guitar!

If you choose Layout from this menu or open the Layout dialog from its button you can select to Display Tablature and also choose Bass, Guitar or MIDI Guitar from the Tablature Preset field; this can be further enhanced (e.g. number of frets available) using the Define option.

Drawing Notes and Editing Them

1. Select the Draw tool ().
2. Select the note value required (). If you can't select the one you want then you may have to change the Display Resolution setting () to a lower value as this governs the lowest value available.
3. Ensure the Snap To resolution is set correctly.
4. Click on the staff at the relevant position to enter the note. You may want to zoom in a little for extra accuracy. The note will sound and the View will update by adding appropriate rests.

5. The note may be edited by dragging with the Select tool () [S] or deleted with the Eraser tool () [E].
6. Right-clicking on a note will invoke a Note Properties dialog which can also be used to edit the note (the Channel field dictates which 'voice' the note belongs in which help when working with multiple voices).

The Staff View also has the ability to be printed so if you have a printer connected you have the usual print capability available including Print Preview and Print Setup from the File menu.

Using the Event List

The Event List shows all events in a text format and is particularly good for tracking down data that isn't easily visible in other Views such as an unusual MIDI controller message.

The View can be opened by clicking on its button in the Views Toolbar () pressing Alt+4 or selecting it from the View menu. Any selected Tracks (MIDI or audio) will show in the list but it also has the same Pick Tracks options as the Piano Roll and Staff View.

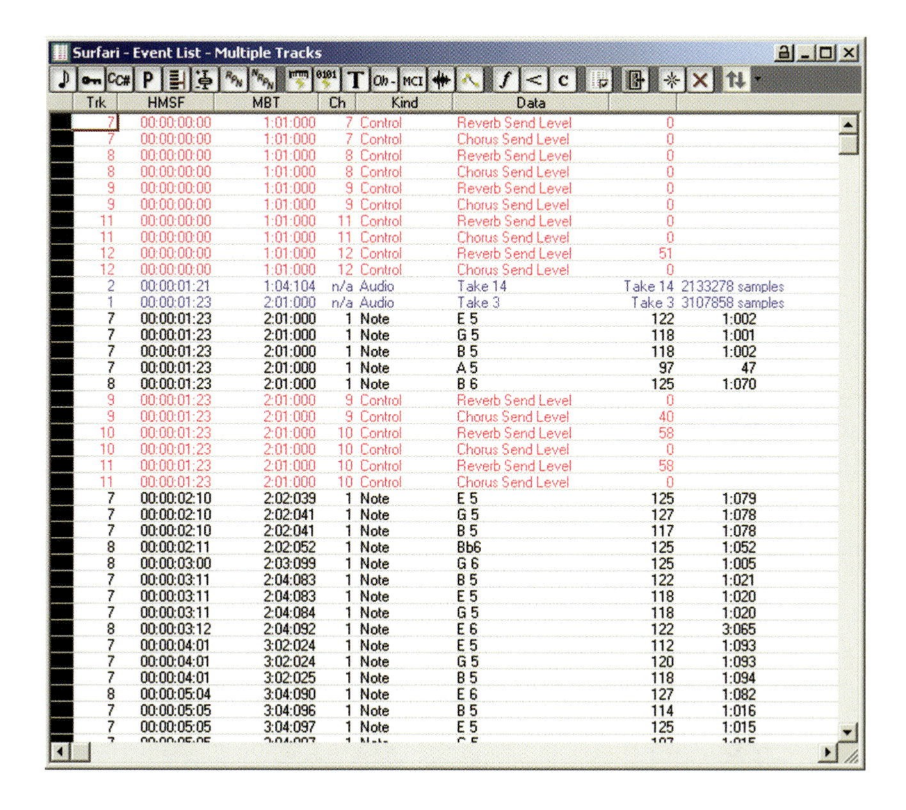

Surfari - Event List - Multiple Tracks

Trk	HMSF	MBT	Ch	Kind	Data		
7	00:00:00:00	1:01:000	7	Control	Reverb Send Level	0	
7	00:00:00:00	1:01:000	7	Control	Chorus Send Level	0	
8	00:00:00:00	1:01:000	8	Control	Reverb Send Level	0	
8	00:00:00:00	1:01:000	8	Control	Chorus Send Level	0	
9	00:00:00:00	1:01:000	9	Control	Reverb Send Level	0	
9	00:00:00:00	1:01:000	9	Control	Chorus Send Level	0	
11	00:00:00:00	1:01:000	11	Control	Reverb Send Level	0	
11	00:00:00:00	1:01:000	11	Control	Chorus Send Level	0	
12	00:00:00:00	1:01:000	12	Control	Reverb Send Level	51	
12	00:00:00:00	1:01:000	12	Control	Chorus Send Level	0	
2	00:00:01:21	1:04:104	n/a	Audio	Take 14	Take 14	2133278 samples
1	00:00:01:23	2:01:000	n/a	Audio	Take 3	Take 3	3107858 samples
7	00:00:01:23	2:01:000	1	Note	E 5	122	1:002
7	00:00:01:23	2:01:000	1	Note	G 5	118	1:001
7	00:00:01:23	2:01:000	1	Note	B 5	118	1:002
7	00:00:01:23	2:01:000	1	Note	A 5	97	47
8	00:00:01:23	2:01:000	1	Note	B 6	125	1:070
9	00:00:01:23	2:01:000	9	Control	Reverb Send Level	0	
9	00:00:01:23	2:01:000	9	Control	Chorus Send Level	40	
10	00:00:01:23	2:01:000	10	Control	Reverb Send Level	58	
10	00:00:01:23	2:01:000	10	Control	Chorus Send Level	0	
11	00:00:01:23	2:01:000	10	Control	Reverb Send Level	58	
11	00:00:01:23	2:01:000	10	Control	Chorus Send Level	0	
7	00:00:02:10	2:02:039	1	Note	E 5	125	1:079
7	00:00:02:10	2:02:041	1	Note	G 5	127	1:078
7	00:00:02:10	2:02:041	1	Note	B 5	117	1:078
8	00:00:02:11	2:02:052	1	Note	Bb6	125	1:052
8	00:00:03:00	2:03:099	1	Note	G 6	125	1:005
7	00:00:03:11	2:04:083	1	Note	B 5	122	1:021
7	00:00:03:11	2:04:083	1	Note	E 5	118	1:020
7	00:00:03:11	2:04:084	1	Note	G 5	118	1:020
8	00:00:03:12	2:04:092	1	Note	E 6	122	3:065
7	00:00:04:01	3:02:024	1	Note	E 5	112	1:093
7	00:00:04:01	3:02:024	1	Note	G 5	120	1:093
7	00:00:04:01	3:02:025	1	Note	B 5	118	1:094
8	00:00:05:04	3:04:090	1	Note	E 6	127	1:082
7	00:00:05:05	3:04:096	1	Note	B 5	114	1:016
7	00:00:05:05	3:04:097	1	Note	E 5	125	1:015

The cursor will be at the same position as the Song Position Pointer (the Now time) and if you click on any of the Event names the cursor will update to that position.

You can hold Ctrl+Shift then press Spacebar to play through each event in steps.

Along the top of the View is a row of buttons that can be switched to hide various Event types, if you click the Note Events button ( ) so that it turns blue then all Note Events will be hidden making it much easier to identify other types of Event.

There is also an Event Manager that can be used to make quick selections. This can be invoked by clicking its button ( ) or pressing [V]. It uses familiar checkboxes to make selections and allows ranges to be selected using the All/None options for MIDI, Special and Notation Events.

Events may also be inserted or deleted by placing the cursor at the appropriate position then using the Insert Event ( ) or Delete Event ( ) buttons. You can also use the Ins and Del keys on your qwerty keyboard.

This View can also be printed.

More MIDI Facilities

SONAR has many MIDI features and facilities that could fill a whole book by themselves, many of which you may never need such as Sysx or creating an Instrument Definition, these are subjects beyond the scope of this book but should you wish to learn more about them, just use the Help files or press F1 for context-sensitive help.

There are some other features that you should really know about as they are very useful but may not be very obvious from a cursory glance. We're going to look at those now starting with . . .

MIDI Drums

Creating drum or percussion tracks using MIDI notes is quite different to using most other instruments and a different focus is required. For instance, a drum hit doesn't need to be shown as a long note but it might require a very large dynamic range to be available in order to express the subtle nuances of playing, hence good control over Velocity is required. To this end Cakewalk have designed a specific drum editing mode available within the Piano Roll that is both easy to use and very flexible.

In order to use the Drum Editor you must have a MIDI Track that has its Output set to a Drum Map. SONAR has several Drum Maps installed by default and you can copy many others from the installation disc to the Drum Maps folder within the SONAR directory should you require them. They cover many pieces of software and hardware, and can save you a lot of work either as they are or as a good starting point when designing your own Maps.

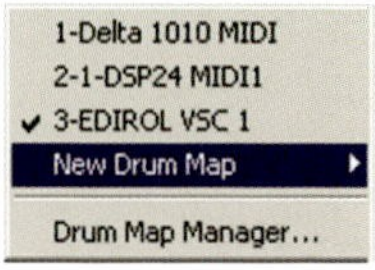

To set the MIDI Output to a Drum Map simply click its Output arrow and select New Drum Map then one of the available Maps.

In order to hear what's going on we need to use a sound source such as a soft synth; so we're going to look at the procedure for drawing and editing in the Drum Editor starting from scratch. If you look on the install CD you may well find a Map for the instrument you are going to use in which case it's a good idea to copy it into your Drum Maps folder before you start as it will be already configured and good to go right away. Look in the Sample Content folder then Drum Maps where you can browse through a group of folders covering various manufacturers and categories.

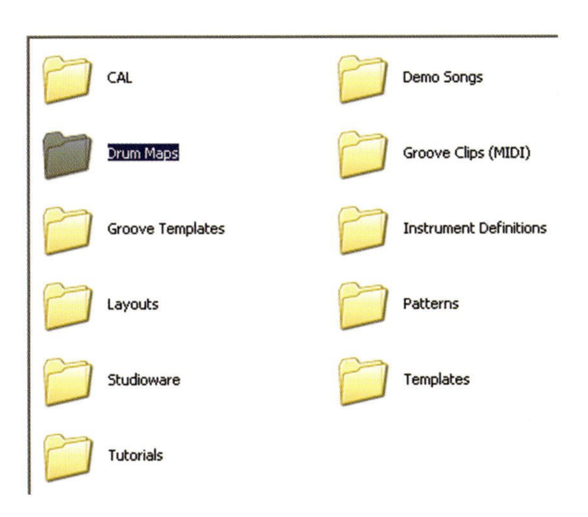

Using the Drum Editor

1. Start with a New Project (a blank is best) and Insert a suitable soft synth by opening the Synth Rack (DXi) and clicking its Insert button (+).
2. Select DXi Synth then choose your weapon! We're going to use the Cakewalk TTS-1.
3. Make sure MIDI Source Track and First Synth Output (Audio) are ticked and click OK, if the synth itself appears just click its (X) button to close it.
4. Open up the MIDI Track and click on the Output arrow then select New Drum Map>GM Drums (Complete Kit).

5 Make sure that the MIDI Track only is selected by clicking on its Track number then open the Piano Roll [Alt+5].

6 The usual keyboard down the left-hand side is now a list of drum notes corresponding to the Drum Map that is loaded. If you click on a note name you should hear it play the corresponding sound from the TTS-1 providing you don't have another MIDI Port or device taking priority. If you don't hear anything then right-click on any name and choose Drum Map Manager . . .

7 In the Manager's Out Port field you need to change all the Ports to Cakewalk TTS-1. Do this by holding down Ctrl+Shift (this changes them all in one go) then clicking the arrow in one of the Out Port fields and changing to Cakewalk TTS-1.

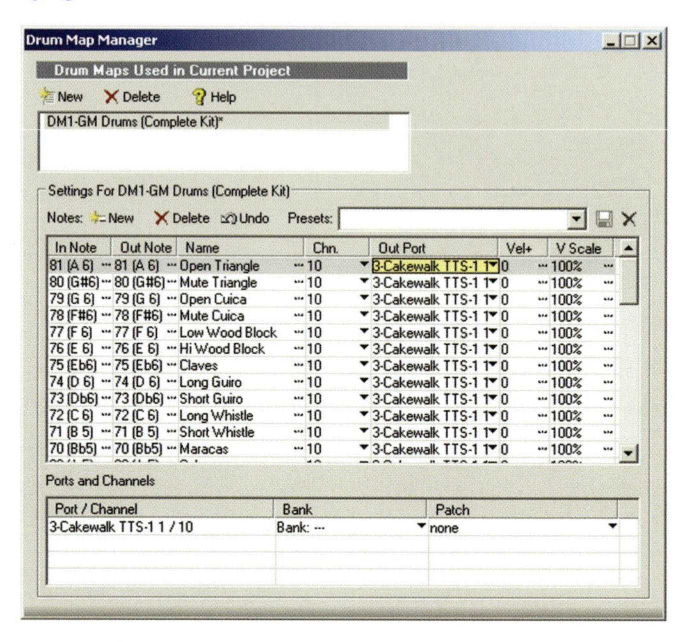

8 Now select the Draw tool and click on the grid to insert a note. It will appear differently to the standard Piano Roll notes as a triangular 'hit'.

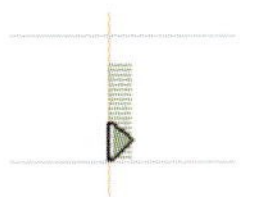

9 Moving the note is simply a matter of placing the cursor over it near the left-hand edge until it becomes horizontal arrows for horizontal movement or to the right when it becomes vertical arrows for moving vertically. There isn't any need to stretch the note so this option isn't available.

10 The Controllers Pane is available as in the standard Piano Roll but there also exists an option to view Velocities for each note by clicking the Show Velocity Tails button ( ) or pressing the Y key.

11 You can now click and drag the note Velocities up or down using the 'ladder' above the note; the value appears as a readout at the upper left of the View. Hold down Shift to change the velocities of all currently selected notes.

12 You may occasionally want to see note lengths, if triggering samples that need to be held for instance, in which case you can just click the Show Durations button ( ) [O] and they will appear.

The same Track selection options are available and it's also possible to work with the Piano Roll View open as well by dragging up the lower horizontal splitter bar. This is particularly useful for working with other rhythm instruments visible such as basslines.

Saving a Drum Map

The Drum Map Manager can be configured to use any available source for any note so you could create a Map that uses kick drum from a soft synth, snare from an external module, hi-hats from a different soft synth, etc., and save it all as a custom Drum Map that can be a real time saver for future Projects.

If you want to change the Properties of a single note then just double-click its name to open its Map Properties page.

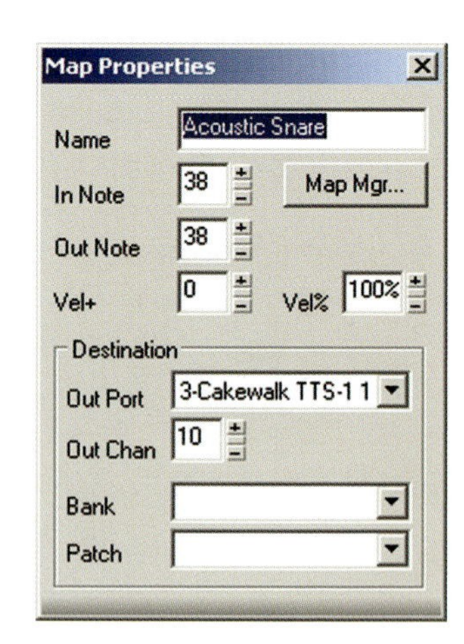

The range of options available means that it's possible to customize Drum Maps so that your hardware Input equipment (e.g. drum pads) can remain as they are but the Outputs can be mapped to different note numbers enabling many variations of instrument to be used without repatching the hardware.

To save your own custom Drum Map simply type a name for it in the Presets field of the Drum Map Manager and click the Save icon. It will then be available in the list of Drum Maps the next time you want to use it.

Using the Pattern Brush

Available from within the Piano Roll and its Drum Editor View, the Pattern Brush is a very useful tool indeed, particularly for writing drum and percussion tracks. Using the Brush means that you can 'paint in' patterns of notes or drum patterns by simply clicking and dragging the tool across the Piano Roll or Drum Editor grid. The tool is mainly aimed at inserting drum patterns but it is also possible to use it for other patterns which can be particularly useful for arpeggio style lines or repetitive drum hits. To do this in the Piano Roll View . . .

1. Make sure that you have a sound source in place such as a soft synth and open the Piano Roll for that Track. If it has a suitable Drum Map then load and configure it, if necessary.
2. Select a note value ( ).
3. Select the Pattern Brush tool by clicking its button ( ) or pressing Q.
4. Click the arrow next to the button and select Note Duration from the menu (if it isn't already selected).

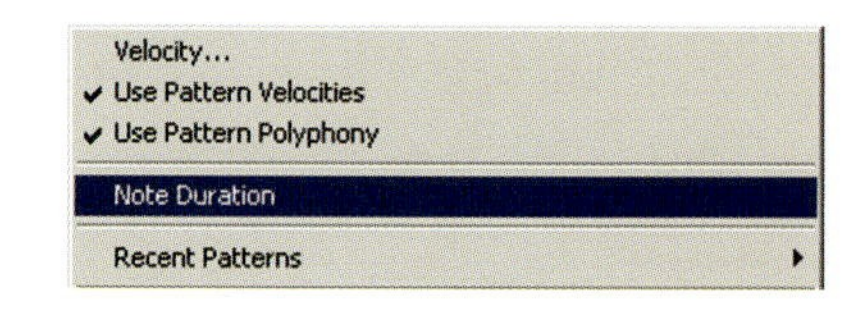

5 Now you can click and drag in the grid area and notes will be inserted at your chosen value.

6 To keep the notes in a single lane just hold down the Shift key while dragging.

7 To specify the Velocity value for the notes to be inserted click the arrow next to the Pattern Brush button again and select Velocity. A dialog box will pop up and allow you to enter a value. Click OK when you're done and all the notes entered will now be at this Velocity.

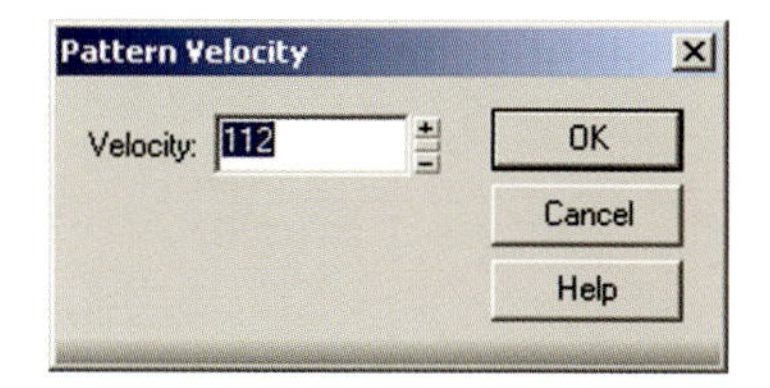

A Template is available to help create drum Tracks using the Pattern Brush. Look in the Quick Start or New Project File dialogs for Pattern Brush Template and follow the instructions when it loads.

Working with Drum Patterns

In order to use the patterns that are supplied with the Pattern Brush you should first have a MIDI drum Track set up and routed to a Drum Map with a source such as a soft synth running from it.

1 Open the Piano Roll for the drum Track which (if everything is set up correctly) will open in its Drum Editor guise.

2 You can check that you have routed things properly by clicking on a note name to hear it play the drum hit that it's assigned to.

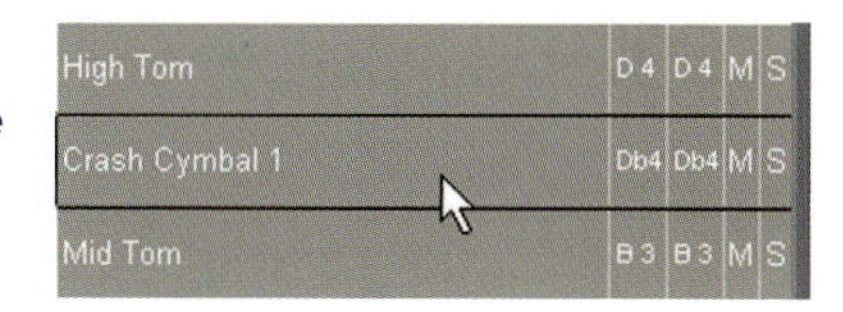

3 Select the Pattern Brush tool by clicking its button ( ) or pressing Q.

4 Click the arrow next to it and select Kick+Snare Patterns (D–F) then Disco 1 from the next menu that appears.

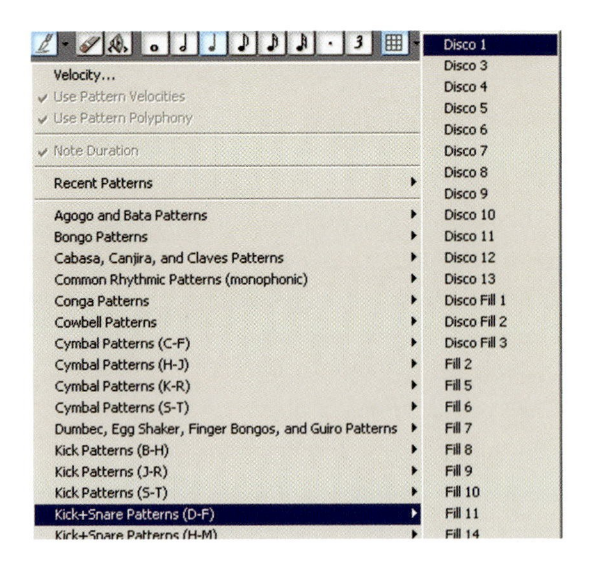

5 Click the arrow to open the menu again and at the top, above the patterns, make sure that Use Pattern Velocities and Use Pattern Polyphony are ticked.

6 You can now click and drag anywhere in the grid and the Brush will paste the selected pattern in, placing the notes in their correct lanes.

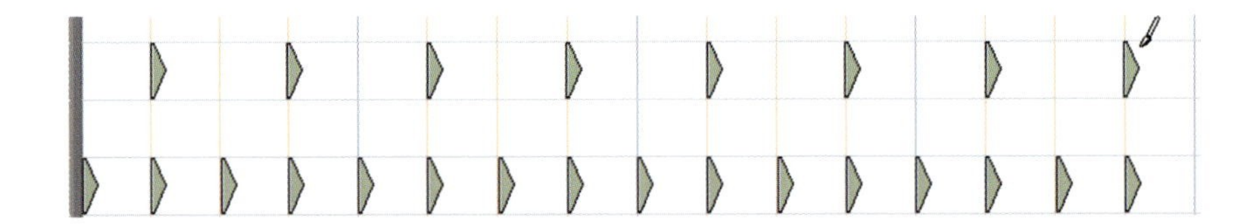

7 Now go back to the patterns and select a Cymbal Pattern to go with your Kick and Snares, one from the Disco selection should work OK, then click and drag in the same place as your previous pattern was entered.

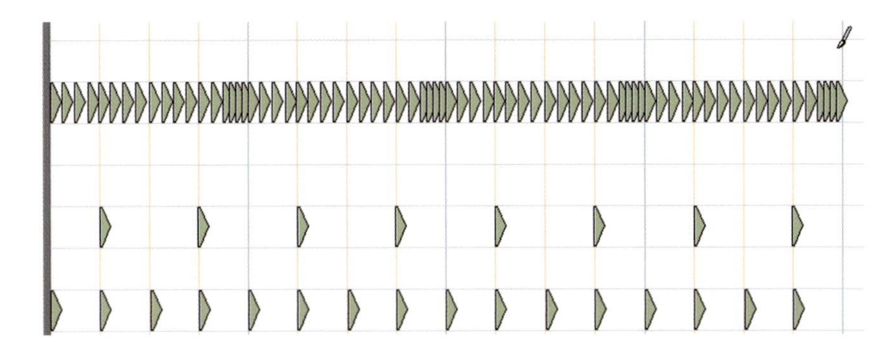

You now have a MIDI drum pattern that can be fully edited as normal. By selecting a variety of patterns and fills it's easy to create full and interesting Tracks without the pain of entering every note individually.

You can also create and save your own patterns for future use or to share like this . . .

1 Create your patterns as a consecutive series using the Drum Editor. They only need to be a single instance of each pattern (a single bar is best).

2 At the start of each pattern insert a Marker by placing the Song Position Pointer at the start of the bar then press F11 (Tight 1 1 Tight 2 2).

3 A dialog box appears where you should enter the name of the pattern and then hit OK.

4 Continue to do this for all of the patterns then immediately after the last one insert a Marker and name it END.

5 In the Track View highlight the MIDI Track by clicking on its number then choose File>Save As.

6 The Save As dialog will open where you can browse to your Pattern Brush Patterns folder (usually in C:/Program Files/Cakewalk/SONAR 4 Producer Edition folder).

7 Name then Save the file as MIDI Format 1 in this folder. If you get a warning about not being able to store Drum Maps in MIDI files then you can just ignore it and click Yes to Save anyway. Next time you run SONAR the patterns will be available and will appear with the names that you gave them.

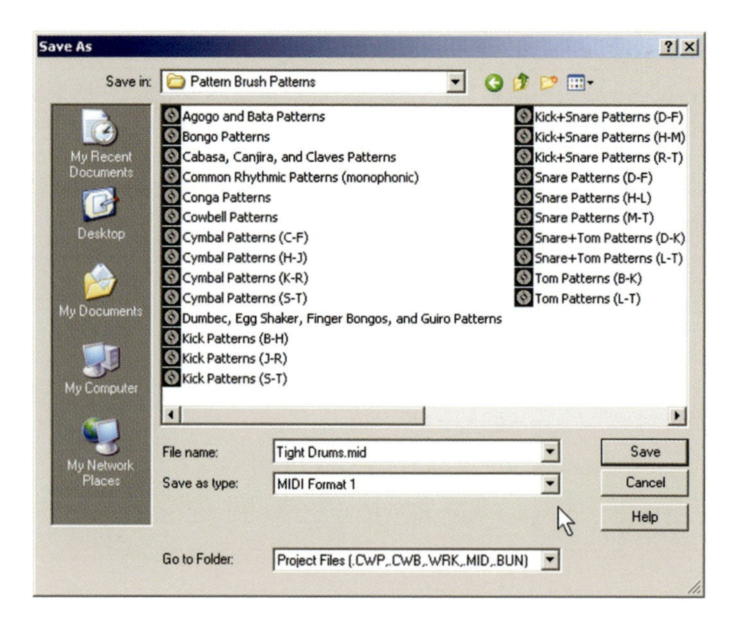
Save As
Save in:
Pattern Brush Patterns
My Recent Documents
Desktop
My Documents
My Computer
My Network Places
Agogo and Bata Patterns
Bongo Patterns
Cabasa, Canjira, and Claves Patterns
Common Rhythmic Patterns (monophonic)
Conga Patterns
Cowbell Patterns
Cymbal Patterns (C-F)
Cymbal Patterns (H-J)
Cymbal Patterns (K-R)
Cymbal Patterns (S-T)
Dumbec, Egg Shaker, Finger Bongos, and Guiro Patterns
Kick Patterns (B-H)
Kick Patterns (J-R)
Kick Patterns (S-T)
Kick+Snare Patterns (D-F)
Kick+Snare Patterns (H-M)
Kick+Snare Patterns (R-T)
Snare Patterns (D-F)
Snare Patterns (H-L)
Snare Patterns (M-T)
Snare+Tom Patterns (D-K)
Snare+Tom Patterns (L-T)
Tom Patterns (B-K)
Tom Patterns (L-T)
File name:
Tight Drums.mid
Save as type:
MIDI Format 1
Save
Cancel
Help
Go to Folder:
Project Files (.CWP,.CWB,.WRK,.MID,.BUN)

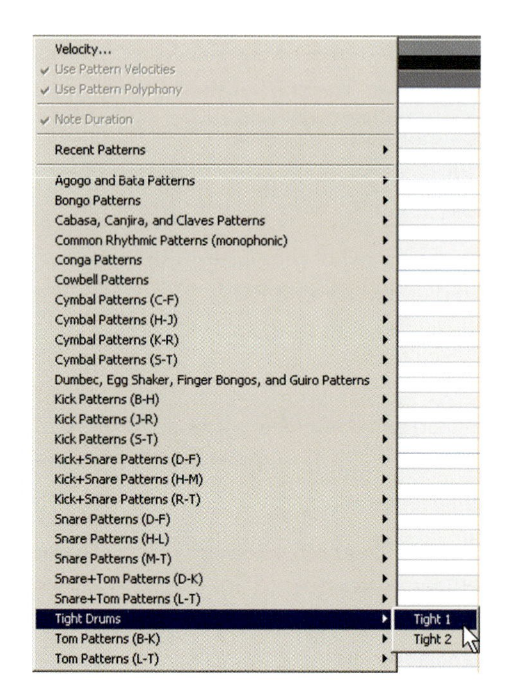
Velocity...
Use Pattern Velocities
Use Pattern Polyphony
Note Duration
Recent Patterns
Agogo and Bata Patterns
Bongo Patterns
Cabasa, Canjira, and Claves Patterns
Common Rhythmic Patterns (monophonic)
Conga Patterns
Cowbell Patterns
Cymbal Patterns (C-F)
Cymbal Patterns (H-J)
Cymbal Patterns (K-R)
Cymbal Patterns (S-T)
Dumbec, Egg Shaker, Finger Bongos, and Guiro Patterns
Kick Patterns (B-H)
Kick Patterns (J-R)
Kick Patterns (S-T)
Kick+Snare Patterns (D-F)
Kick+Snare Patterns (H-M)
Kick+Snare Patterns (R-T)
Snare Patterns (D-F)
Snare Patterns (H-L)
Snare Patterns (M-T)
Snare+Tom Patterns (D-K)
Snare+Tom Patterns (L-T)
Tight Drums
Tom Patterns (B-K)
Tom Patterns (L-T)
Tight 1
Tight 2

Adding Autocue Style Lyrics

There are two Views that allow the insertion of lyrics, either the Lyrics View or the Staff View. If you want to read the lyrics as the song plays in the style of an Autocue or Karaoke then the Lyrics View is better as the Font and its size are editable to suit your requirements so you can make them very large and easily visible, the words are also highlighted as the song plays.

To Enter Lyrics in the Lyrics View

You must have a suitable MIDI Track within your Project preferably one that has notes corresponding to the melody line. If you haven't then it's worth creating one. It doesn't have to be audible as the notes are just there for the lyrics to 'adhere' to.

1. Select your melody Track and then open the Lyrics View using its button (L) selecting it from the View menu or pressing Alt+8.
2. You'll see a series of hyphens which correspond to the notes on your Track. Type your lyrics into the pane to correspond with the song and its notes. If it helps you can resize the Views so that you can see the Track View or any other View as well.
3. If you need to extend the lyrics over more than one note you can add hyphens [-] between syllables such as 'par-ty time'.

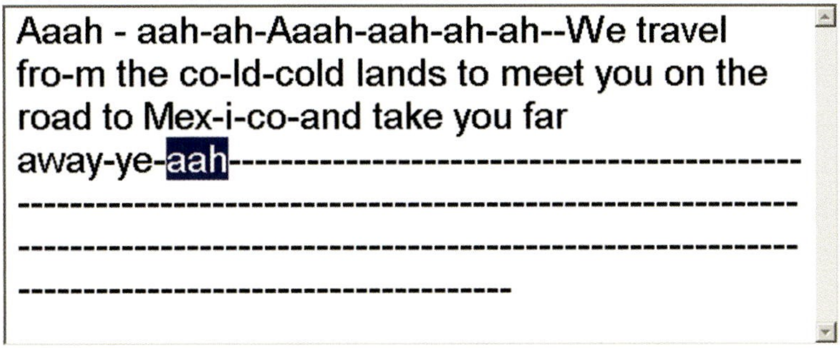

Aaah - aah-ah-Aaah-aah-ah-ah--We travel
fro-m the co-ld-cold lands to meet you on the
road to Mex-i-co-and take you far
away-ye-aah--

----

4. When you play back the song your lyrics will be highlighted as it plays and you can easily edit any errors. Like most things it all becomes clear after a couple of attempts!

5 Use the Font . . . button (Font...) to open the font options dialog box. You can also set up two different fonts that can be quickly changed by using the f_a and f_b buttons (f_a f_b). Select one and click the Font · · · button to change its properties.

To Enter Lyrics in the Staff View

1 Select a suitable melody MIDI Track and open the Staff View using its button, the View menu or press Alt+7.

2 Click the Lyrics button (L) or press Y.

3 Place the cursor just underneath a suitable note so that it shows the Draw tool then click, a text box will open.

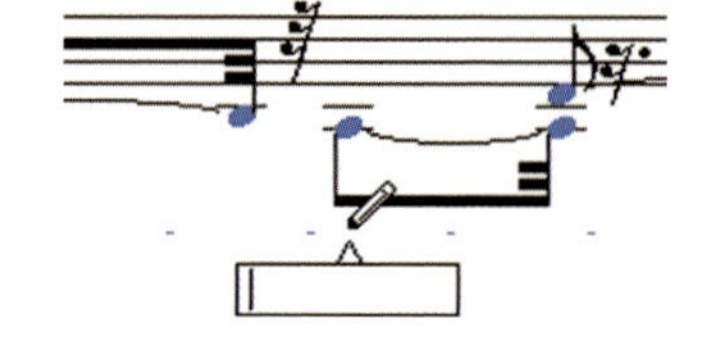

4 Type in your lyrics pressing the Spacebar to move on to the next word.

5 If you need to extend a word over more than one note just insert a hyphen between the syllables [-].

6 To edit a word just click on it with the Draw tool to open the typing box again.

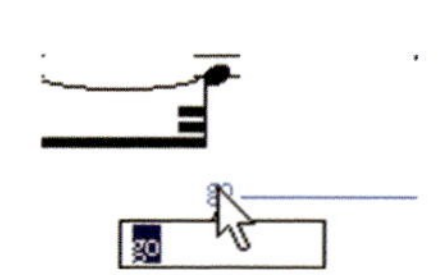

Lyrics entered here are also visible by selecting the Track and opening the Lyrics View so if you prefer to enter lyrics in the Staff View you can still use the Lyrics View for playback.

CHAPTER 7
USING PLUG-IN INSTRUMENTS AND EFFECTS

Plug-in instruments and effects have changed the face of modern music composition and in many cases have negated the need for hardware so they are vitally important. SONAR handles their integration very elegantly and logically keeping them user-friendly but still immensely powerful.

What are Plug-ins?

A Plug-in is what it sounds like, a software processor that is 'plugged in' to the main sequencer system at an appropriate point.

If you're familiar with mixing consoles you may well know about insert points and send-return loops that are used for effects and processors. Plug-in effects work in the same way. Here's the Cakewalk Sonitus:fx Reverb plug-in.

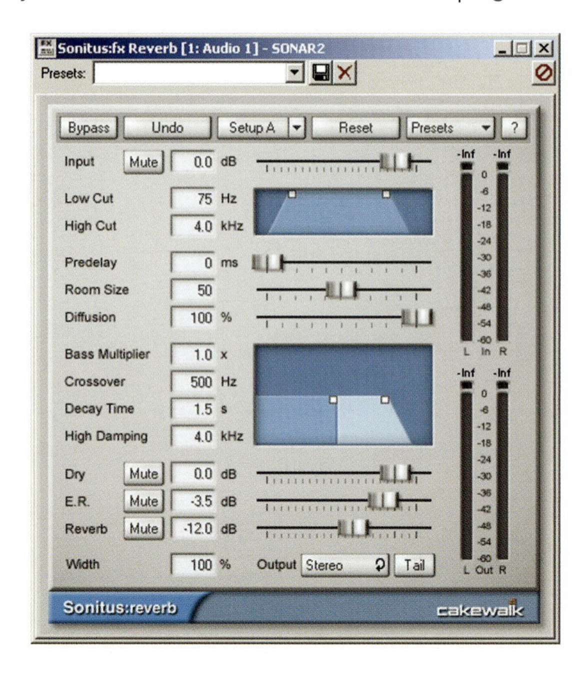

Plug-in instruments, or soft synths, are usually inserted within the MIDI Input chain and feed their output signal to an audio channel so that if you play a MIDI note into the instrument it will process the data and generate a relevant audio sound that is passed to its output. There are a few exceptions but we'll stick with this description as it covers the large majority of current soft synths.

Here's the Cakewalk TTS-1 soft synth.

MIDI Effects

MIDI Tracks may use special MIDI effects plug-ins in a similar fashion to audio effects by simply right-clicking in a MIDI Track's FX Bin and choosing an effect from the MIDI Effects menu that appears.

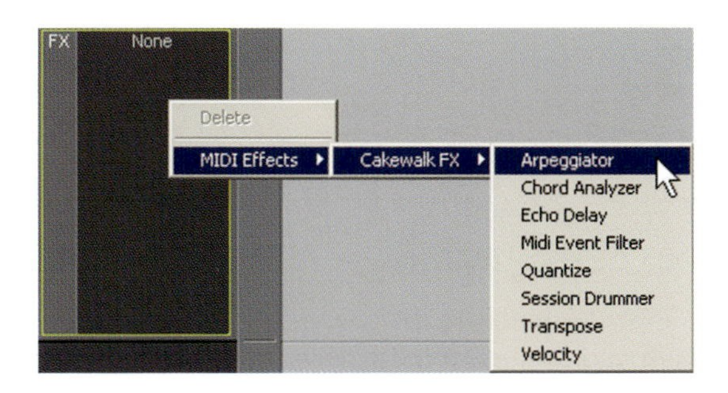

These effects don't have an activation button to turn them on and off but they can be edited by double-clicking on their name in the FX Bin. The plug-in will then open and can be tweaked accordingly.

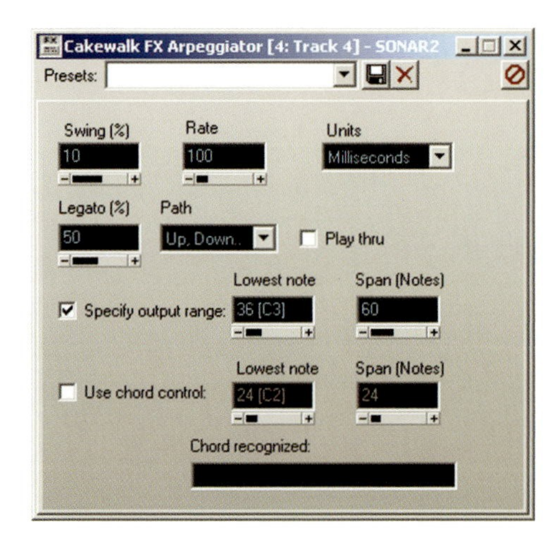

It's worth taking a look at the effects you have available as they can be very useful in a variety of circumstances, for instance using a Transposition plug-in could make life easier when you have a MIDI Track routed to a synth that has lots of kick drum samples, each one on a different note. You could play the song and work through each kick drum sound until you found the one that works best by simply transposing one step at a time. If you change your mind you haven't done anything destructively so it's easy to change back.

VST or DX?

You may have heard about various types of plug-in standard such as VST or DX and a lot has been said about the merits of each one but in general there isn't any real difference to the user providing the software has been written well. Some plug-ins may be available in either format in which case I'd recommend trying the DX version first simply because it is Cakewalk's native format but either version should work just as well. If one works better than the other then simply use that one!

DX plug-ins will usually come with a built-in installer or specific instructions to install them so adding any new ones to your system is simply a matter of following the included documentation.

SONAR has a special system for registering VST plug-ins using the VST Adapter utility which will be run during your first install. Whenever you want to add a new VST plug-in follow this procedure . . .

1. Unless the VST plug-in has its own installer it will usually consist of a .dll file and possibly a manual or Help file. The .dll file should be placed in your VST folder. If you're not sure where it is (typically C:/Program Files/Steinberg/Vstplugins or similar) then use the Windows Search facility to find a folder with VST in its name. You can have more than one folder providing you 'tell' the Adapter where they all are.

2. Open the VST Adapter from the Start> All Programs>Cakewalk>Cakewalk VST Adapter menu.

3. Click Next then make sure your VST folder paths are shown, if not use the Add button to add other paths. Click Next and the Adapter will detect your plug-ins and show them in the next screen.

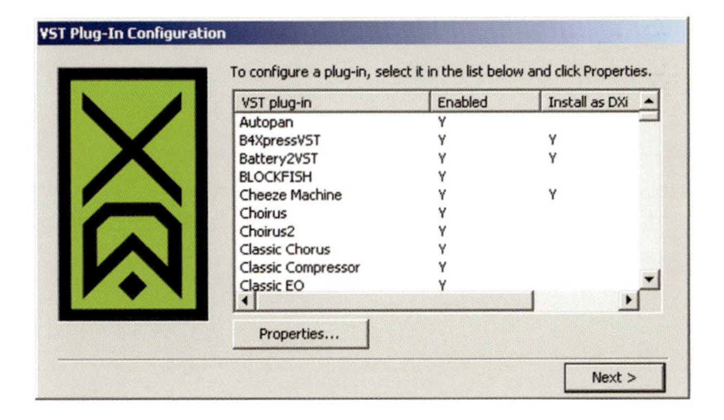

4. From the Plug-in Configuration screen you can highlight a plug-in and press Properties to change its properties if desired. When you've finished click Next.

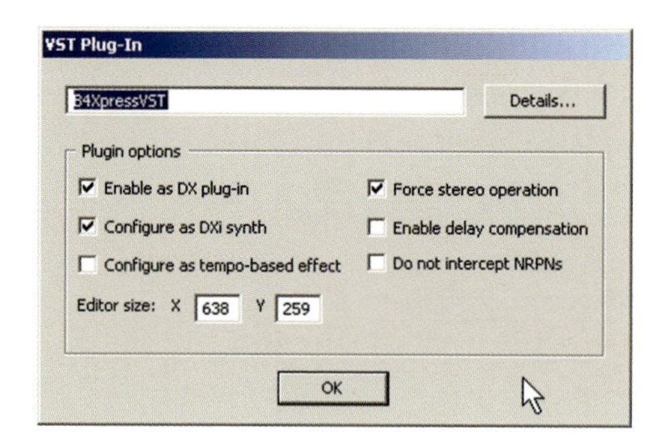

5 The Adapter will configure all of your plug-ins. Click Finish to exit.

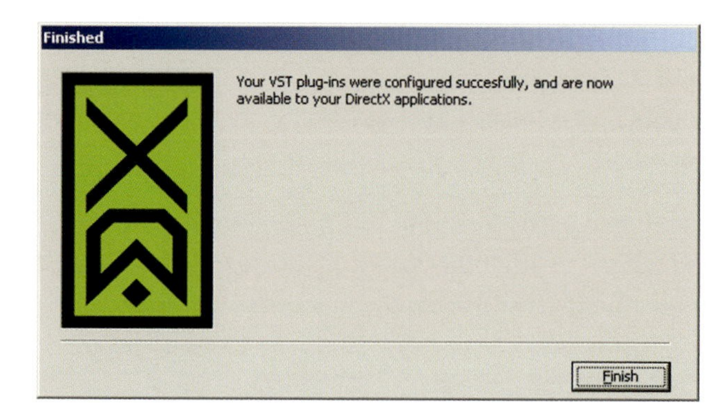

The Synth Rack

All plug-in instruments are usually loaded using the Synth Rack regardless of whether they are DXi or VSTi, (although they can be inserted in audio FX Bins) here's how . . .

1 To open the Synth Rack simply click its button ( ). The Rack can be resized to accommodate more instruments if required by dragging its edges.

2 Click the Insert button in the Rack or press A to see a menu appear. You'll see an option for DXi Synth

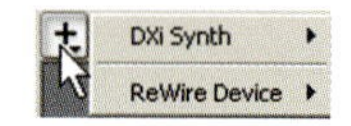

(which includes VST's) if you have any ReWire devices you'll also see an option for them but ignore it for now and follow the DXi Synth link and click on your chosen Synth. Note that you may have further submenu's depending on what is installed on your system.

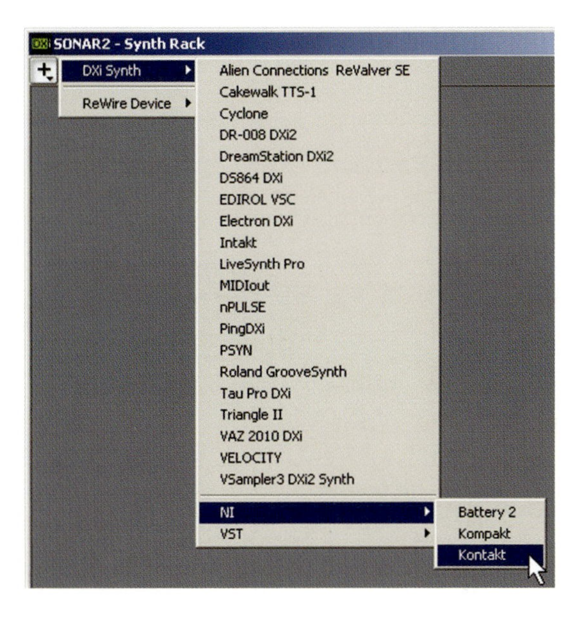

3 A Synth Options dialog now allows you to choose a variety of properties for the Synth as follows:

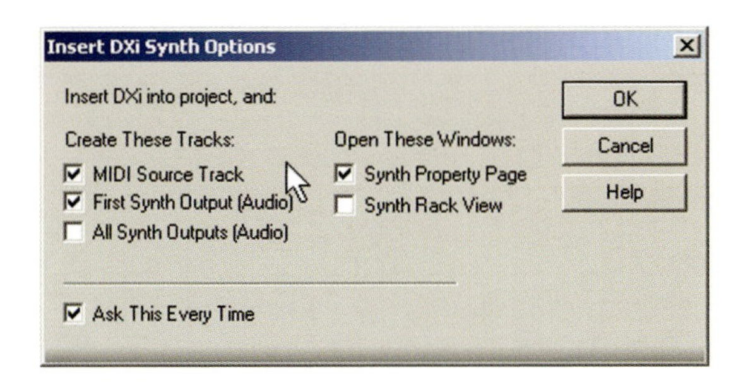

MIDI Source Track will insert a MIDI Track already configured and routed into the synth so it's usually required.

First Synth Output (Audio) will insert a stereo audio Track already configured and routed from the synth's first stereo Output pair.

All Synth Outputs (Audio) will insert multiple audio Tracks already configured and routed from the synth's Outputs. The number will depend on the synth itself but be careful as some instruments may load 64!

Synth Property Page simply shows the synth interface.

Synth Rack View shows the Synth Rack when a synth is inserted.

If you don't see the synth interface it can be recalled by double-clicking on its strip in the Synth Rack or highlighting it and pressing the Properties button or P key.

Clicking the (X) button will delete the currently selected synth but not the associated Tracks.

Inputs to Your Instrument

You can add further Input Tracks to the instrument by inserting a new MIDI Track into your Project then selecting the instrument from its Output field.

The instrument's sounds can be selected by using the Bank and Patch fields in the MIDI Track.

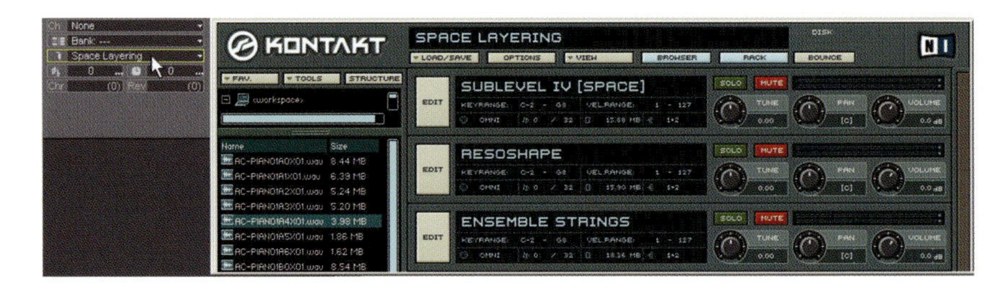

Although the incoming MIDI signal is usually routed to whichever Track is currently in focus it's possible to play more than one instrument at the same time using two or more MIDI input devices (such as drum pads and a keyboard) by selecting a different MIDI Channel for each Hardware device and a corresponding one on each soft synth.

1. Set up an extra soft synth with a MIDI Track routed to its input so that you have two soft synths running complete with MIDI and audio Tracks.

2. Configure the MIDI Tracks in SONAR to match the incoming Channel numbers of the hardware devices (e.g. one and two as set on the hardware Transmit Channels) and select suitable Banks and Patches on each synth.

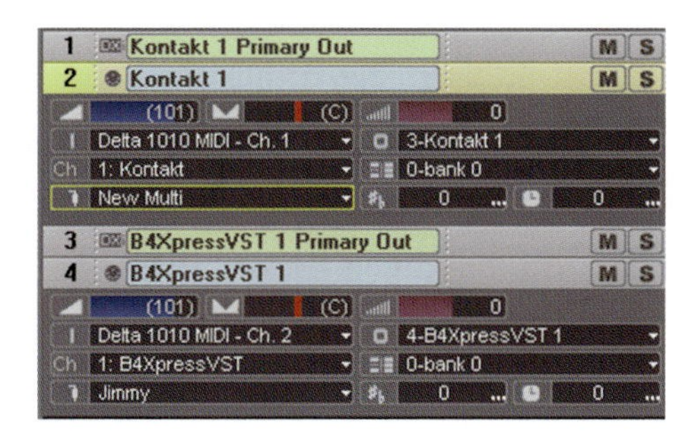

3. Switch the Input Echo buttons ( ) of both Tracks to On.

4. You should now be able to play both instruments and hear their assigned sounds.

Manipulating the Output Signal

Because the soft synth Outputs are sent through 'normal' audio channels they can be treated as any other audio channel or Track so you can add effects, Automation and also route the signal through other Buses in any way you like. Another feature that is available when multiple outputs are used is the ability to send different sounds from the same instrument to different outputs thus allowing you to effect each sound in a different way, maybe some reverb on drums and chorus on the bass for instance.

The ability to do this will depend on the instrument itself but if it has multiple outputs then you can add an audio Track and assign its input to one of the

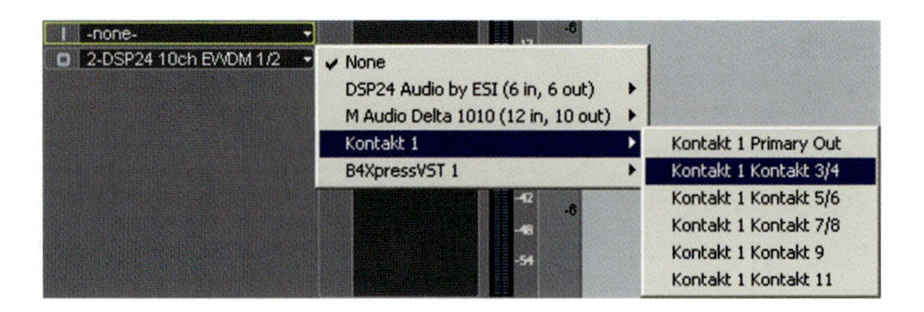

instrument's outputs. Whatever sound is routed to that output will then appear through the new audio channel and can be effected or routed like any normal audio channel.

Most soft synths will have specific routing options, here's how to route to different audio Outputs using the Cakewalk TTS-1:

1. Use the Synth Rack to load up the TTS-1 and choose All Synth Outputs (Audio) from the Options dialog (All Synth Outputs (Audio)).
2. From the synth's main interface, click the System button on the right (SYSTEM).
3. The System Settings dialog will appear, now click OPTION.
4. The Output Assign tab should have the Use Multiple Outputs box checked and all parts will have box 1 checked in the OUTPUT column.

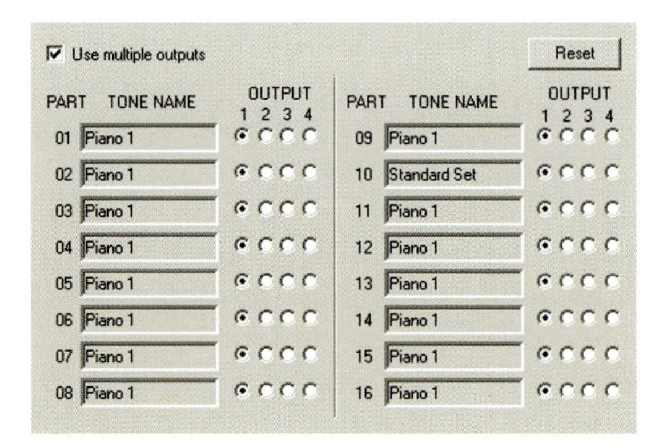

5 To send a part through a different audio output simply check the number next to the TONE NAME field that you want it to be routed to.

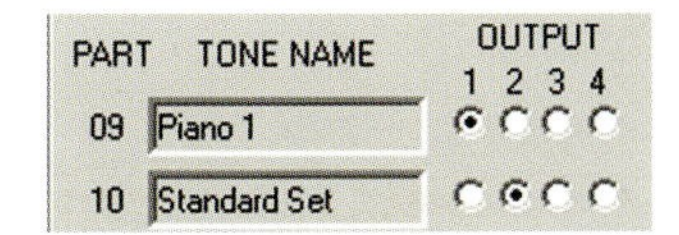

6 After closing the dialogs you'll see that when played each part will appear from the output you chose allowing some flexibility for mixing and effecting. Here we have the piano coming from Output 1 and the drums coming from Output 2.

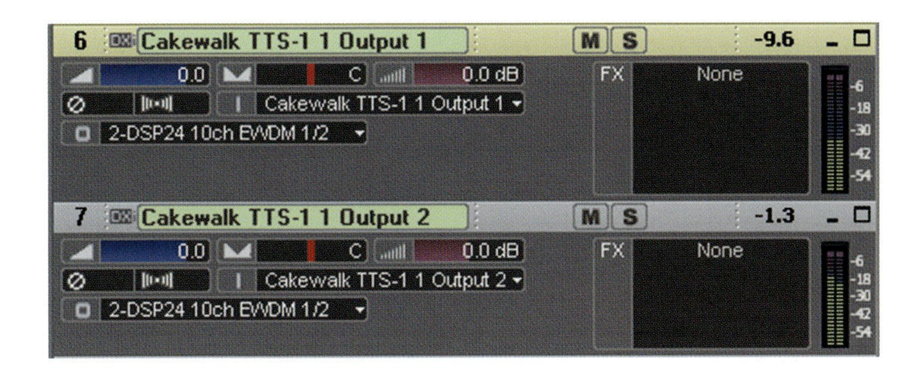

How Can I Deal with Latency?

When playing soft synths in real-time you may experience latency which will be felt as a delay between the note being played and its sound being heard. This is basically due to the MIDI signal having to work its way through your system, get processed and then come back out of the other side as audio. Your soundcard and its drivers will be the main cause of this delay so using a good quality soundcard and keeping the drivers up to date is essential. Whenever you change soundcards or update the driver be sure to run the Wave Profiler by going to the Options>Audio page and clicking the Wave Profiler button.

Latency can also be adjusted here by dragging the Mixing Latency slider to the left to make the response faster or to the right towards Safe to give the system a bit of processing headroom back. In my experience it's a good idea to keep it a couple of notches up from the Fast setting unless you need to play a soft synth in real-time. If you experience any stuttering or dropouts then it's often curable by dragging the slider a little to the right.

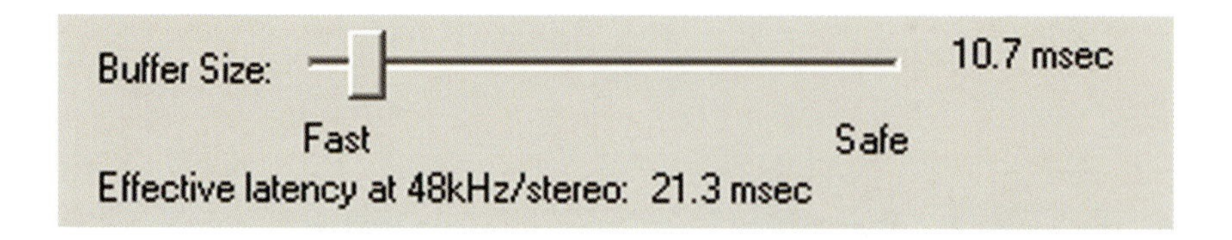

It is possible to lower the latency figures by changing some of the other settings such as the Buffers in Playback Queue but always make a note of your old setting in case you screw things up!

Plug-in Effects

Plug-in effects can replace a whole rack full of hardware and there are a host of them integrated into SONAR. An integral part of any studio this chapter will explain the principles of their use within the SONAR package alongside the plug-in instruments which have some similarities.

Inserting a Plug-in Effect

To use an audio effect plug-in is very easy, all you have to do is right-click in the FX Bin of an audio Track then select one from the list of Audio Effects that appears.

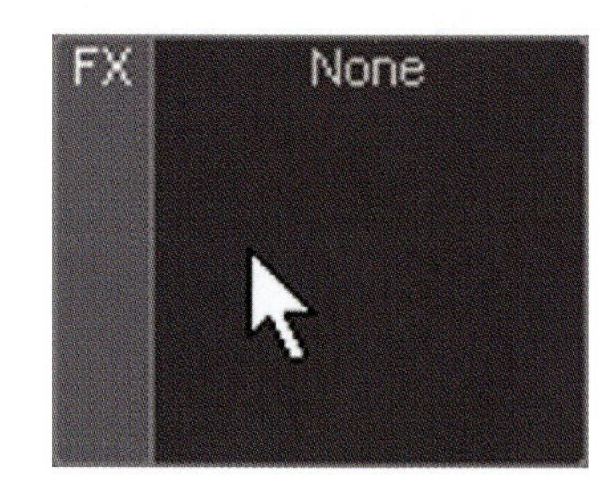

It will be inserted and its interface will open.

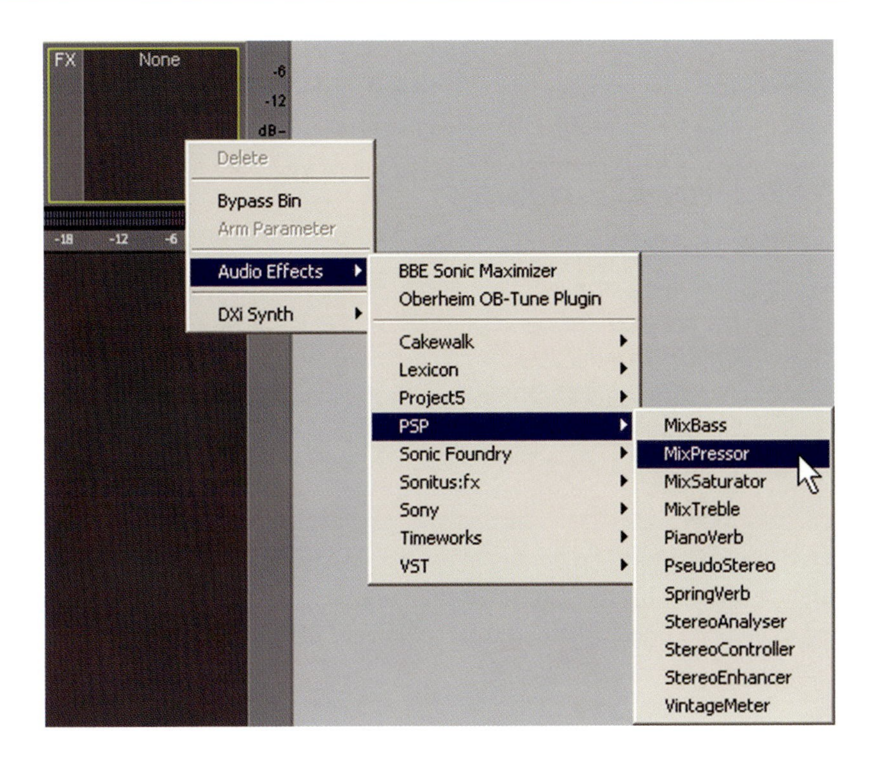
FX
None
Delete
Bypass Bin
Arm Parameter
Audio Effects
DXi Synth
BBE Sonic Maximizer
Oberheim OB-Tune Plugin
Cakewalk
Lexicon
Project5
PSP
Sonic Foundry
Sonitus:fx
Sony
Timeworks
VST
MixBass
MixPressor
MixSaturator
MixTreble
PianoVerb
PseudoStereo
SpringVerb
StereoAnalyser
StereoController
StereoEnhancer
VintageMeter

PSP MixPressor [1: Audio 1] - SONAR3
Presets:
EDIT
universal
compress
slope
PM VU GR
LEFT
PRE
POST
RIGHT
20%
DEL
SCL
PROCESS
RMS
LIM
80%
input
freq
Q
attack
hold
release
make-up
output
0.0dB
612Hz
0.0
fast
0.00ms
slow
auto
0.0dB
PSP MixPressor
mix
100%

You can do this in either the Track View or the Console View.

Inserting effects into Buses is exactly the same.

Live Real-time Effects Plug-ins

If you have a fast system and you can achieve low latency figures then you should be able to use real-time effect plug-ins. Just insert the plug-in and then click the Input Echo button ( ) so that it's On but it's a good idea to turn your speakers down first just in case the level is high enough to create feedback. You'll hear the effect as if it was an effect module or pedal when playing an instrument that's routed into this audio channel.

If you record an audio Track with an effect patched in like this you will only record the 'dry' audio and not the effect. You can of course apply the effect destructively later by using the Process>Apply Audio Effects command but as long as you don't perform this operation you still have the option of changing it or removing it completely.

Where to Use Dynamic Effects

Dynamic effects such as compressors and equalizers are most often used as Track Inserts directly in their FX Bins or in Buses that control Subgroups such as drums. SONAR's flexible Bussing structure allows you to place them anywhere in the signal chain.

Where to Use Creative Effects

Creative effects such as reverb and chorus are often placed directly in Track FX Bins but are also useful when shared between several or indeed all Tracks by placing them in their own Bus. Tracks can then have a Send to the Bus which can be switched either Pre or Post Fade and have its own level and

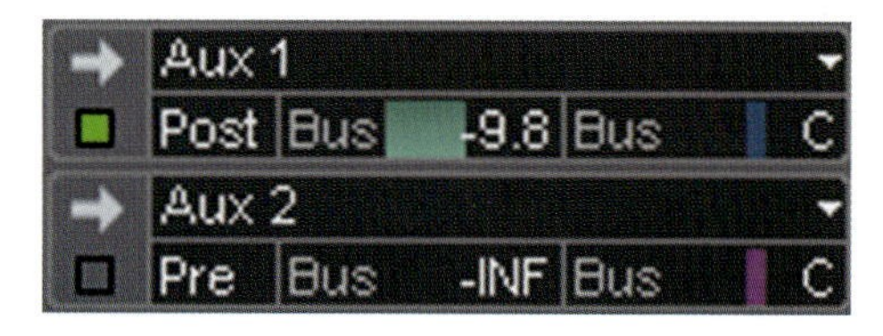

pan going out to the Bus providing a balance between the 'dry' and 'effected' signals appearing at the final output.

Tracks can also be routed directly into Buses sending the full signal, this is usually how a Subgroup is set up.

Equalization in SONAR

EQ plug-ins can be inserted anywhere there is an FX Bin the same as any other effect plug-in but SONAR Producer edition has built-in EQ facilities which utilize plug-ins that you don't have to insert yourself. To activate the EQ on a Track . . .

1 Select the Track by clicking its number (1).

2 Ensure the Inspector or channel strip is visible on the left of the Track View. If it isn't, click the Inspector button or press the I key.

3 Make sure that EQ is active in the strip by clicking this button which has two modes.

4 When the button is green you can activate the EQ by clicking the EQ Enable/Disable button and then either select a band enable it and choose a filter type and drag its sliders to adjust or . . .

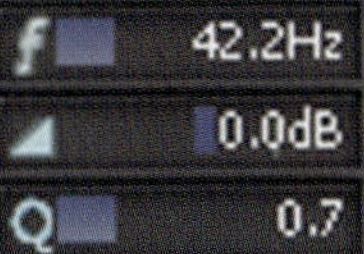

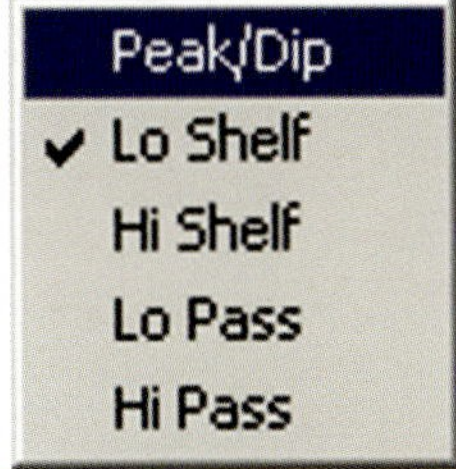

5 Double-click the 'Plot' to open the full Equalizer plug-in. Each band can then be activated by clicking its switch and adjusting its parameters by typing in new values, selecting a filter shape from the five available and dragging the level slider.

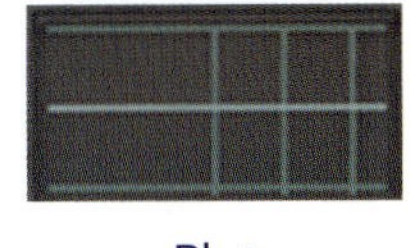

Plot

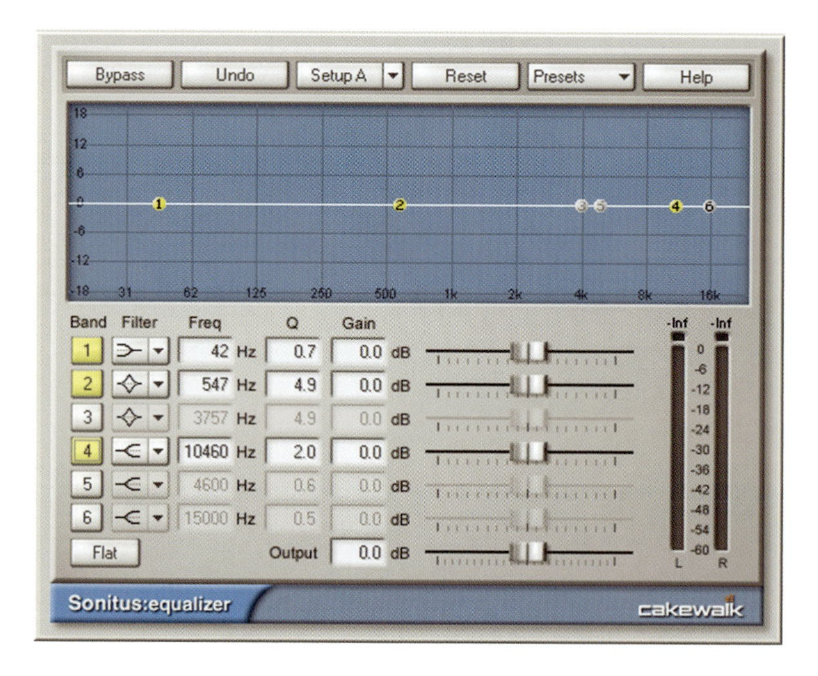
Bypass
Undo
Setup A
Reset
Presets
Help
18
12
6
0
-6
-12
-18
31
62
125
250
500
1k
2k
4k
8k
16k
Band
Filter
Freq
Q
Gain
1
42 Hz
0.7
0.0 dB
2
547 Hz
4.9
0.0 dB
3
3757 Hz
4.9
0.0 dB
4
10460 Hz
2.0
0.0 dB
5
4600 Hz
0.6
0.0 dB
6
15000 Hz
0.5
0.0 dB
Flat
Output
0.0 dB
-Inf
-Inf
0
-6
-12
-18
-24
-30
-36
-42
-48
-54
-60
L
R
Sonitus:equalizer
cakewalk

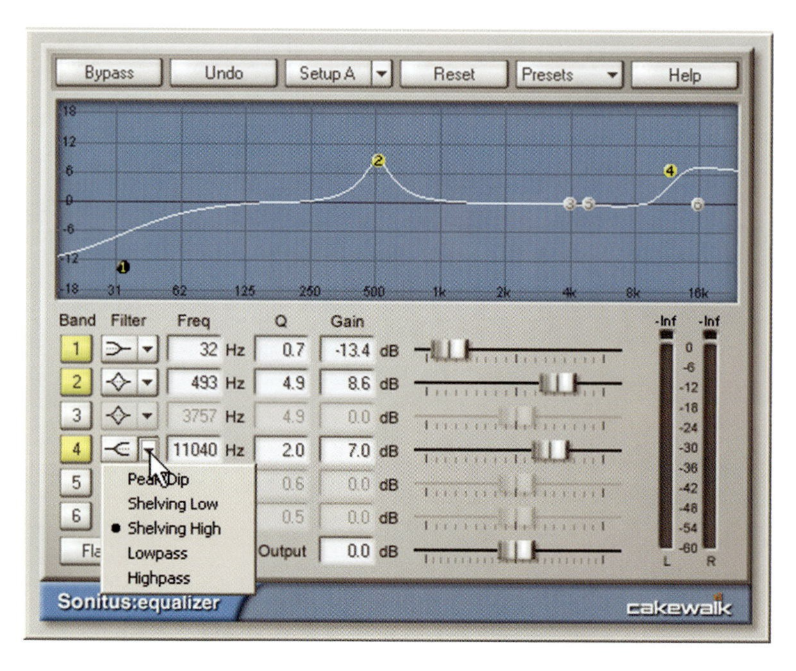
Bypass
Undo
Setup A
Reset
Presets
Help
18
12
6
0
-6
-12
-18
31
62
125
250
500
1k
2k
4k
8k
16k
Band
Filter
Freq
Q
Gain
1
32 Hz
0.7
-13.4 dB
2
493 Hz
4.9
8.6 dB
3
3757 Hz
4.9
0.0 dB
4
11040 Hz
2.0
7.0 dB
5
0.6
0.0 dB
6
0.5
0.0 dB
Peak/Dip
Shelving Low
Shelving High
Lowpass
Highpass
Output
0.0 dB
-Inf
-Inf
0
-6
-12
-18
-24
-30
-36
-42
-48
-54
-60
L
R
Sonitus:equalizer
cakewalk

6 If you click the EQ button ( ) again so that it turns blue you'll have access to a different mode of operation where the Equalizer settings are available as a series of sliders that can be activated by clicking the empty square button for each slider.

7 The Filter type can be selected from the menu that appears when the bar is clicked. It has the same five options as the plug-in interface.

8 Frequency, Gain and Q may be set by dragging their corresponding sliders. Note that only the first four bands are available this way, you'll have to open the plug-in to access the other two.

Stacking and Chaining Effects

If you've used many hardware effects before you'll probably know that it's possible to connect them together in a chain feeding one effect's output into the input of another. Doing this is fine but bear in mind that you can generate a different sound by altering the position of each effect. For instance if you place a compressor on a vocal then put a reverb after it you'll be compressing the vocal and adding reverb to the compressed signal, but if you place the compressor after the reverb you'll also compress the reverb which will make its effect become more pronounced in relation to the vocal. Both are valid methods and SONAR makes it easy to try out such scenarios by simply dragging the effects into a new order:

1 Fire up a suitable audio Track and insert an effect plug-in into its FX Bin by right-clicking on the Bin and choosing from the list. Play the Track then choose a suitable preset on the effect or tweak it until it sounds OK then close its interface.

2 Click the green box to turn off the effect, the box will clear (Reverb).

3 Insert another different effect in the same way and tweak it until it sounds good with your source material then close its interface.

4 Now click the switch back on for your first effect to hear them both working together.

5 With the Track playing you can now click on one of the effects in the FX Bin and drag it up or down to change its position in the chain. You'll hear a difference in the sound depending on where each effect is placed. You don't have to stop at two though as you can add many more although it probably won't sound too good!

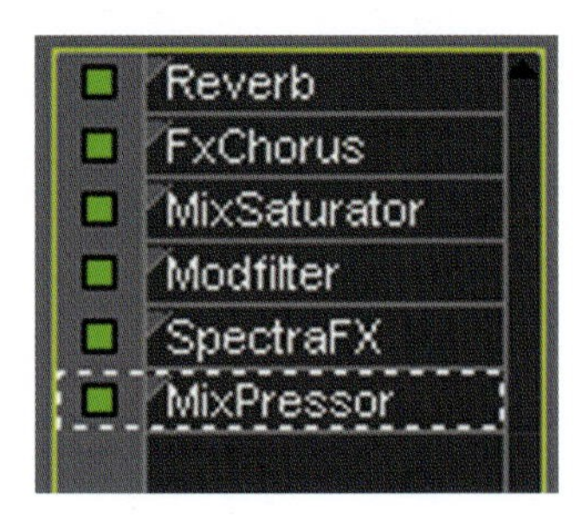

6 To quickly hear the 'dry' sound again just right-click on the FX Bin and select Bypass Bin. Repeat the procedure to reinstate it.

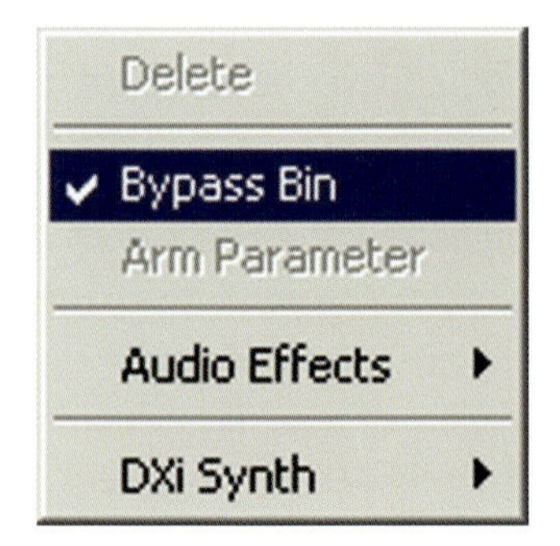

Effecting Several Channels

There are many occasions when the same effect may be required on more than one Track either at the same level, in a Subgroup for instance, or with varying amounts of the effect such as an overall ambient reverb where instruments appear to be at various distances within the same space.

To use an effect over a group of Tracks in the form of a Subgroup you'll need a Project with several suitable Tracks to try this out on such as an ensemble or multi-tracked drums, then . . .

1 Open the Bus pane at the lower section of the Track View either by dragging the horizontal splitter bar upwards or by pressing the show/Hide Bus Pane button ( ). Resize it to provide a little working area.

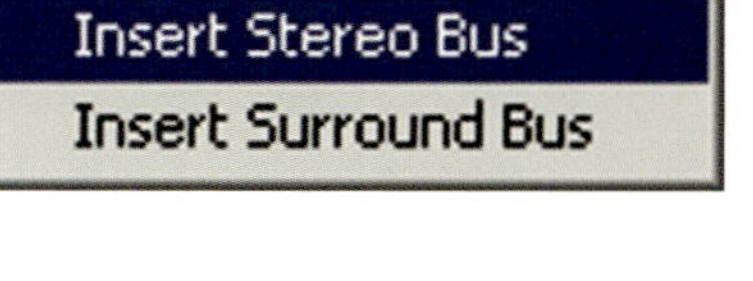

2 Right-click in a blank area of the Bus pane and click Insert Stereo Bus.

3 Click the Restore Strip Size button ( ) or drag the new Bus panel open.

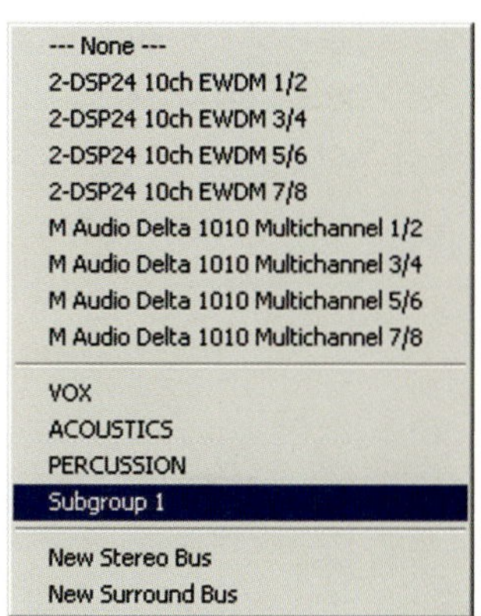

4 Right-click in its FX Bin and insert an appropriate effect.

5 Name the Bus by double-clicking its name strip and typing something appropriate like 'Subgroup 1'.

6 Go to the Tracks that are to be routed into this Subgroup and change their outputs to the Bus that you just named by clicking on the arrow in their output ( ) field and selecting it from the list.

7 Press Play (Spacebar) and you'll hear the effect on all the routed Tracks. In order to adjust it you simply open up the effect by double-clicking on its name in the Bus and then you can tweak it. Most often you'll need to alter the Mix parameter or adjust the dry/wet effect levels.

To use an effect on a Send or Auxilliary Bus . . .

1 Insert a new Bus as above and insert your chosen effect.

2 Name the Bus something appropriate like Aux Bus 1.

3 Right-click in the control panel of a Track that is to be sent to the effect and select Insert Send then choose the new Bus by the name you just entered. (You can select more than one Track and Insert Sends to all of them simultaneously if you want.)

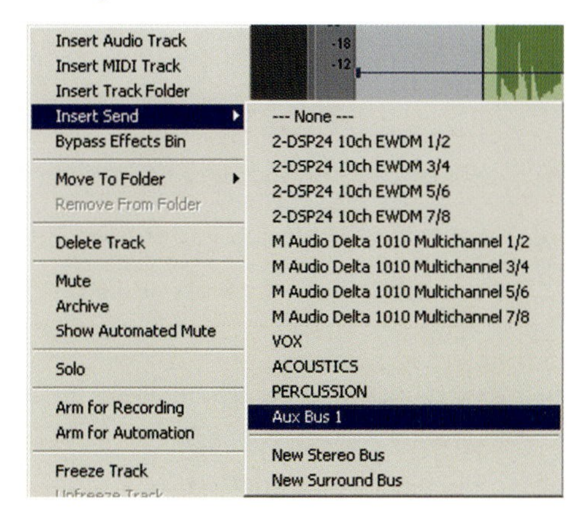

4 A new Send control will appear that can be activated by clicking the box at the left ( ).

5 Clicking Post will change the Send to Pre or Post fader and the Level and Pan can be adjusted using the sliders ( ).

6 Clicking the box above them will allow you to reassign or delete the Send.

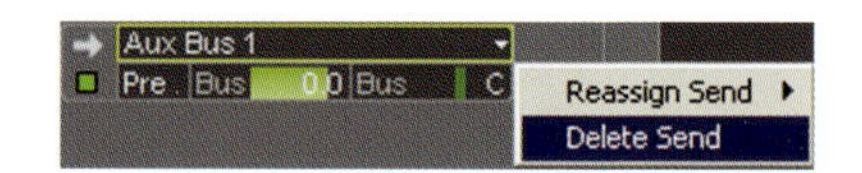

7 Play the Track and adjust the Send controls to suit, you may also have to set the effect Mix quite high or fully up (wet) to achieve proper Send levels.

8 The Bus parameters can also be adjusted as required.

Surround Effects

If you're lucky enough to have a surround sound system then you'll be pleased to know that apart from dedicated surround effects SONAR also has the ability to implement 'normal' effect plug-ins as surround effects. It does this by adding enough instances of the plug-in to cover all of the necessary outputs but allows you to control all of the instances from just one interface.

The plug-in's interface will be shown with several tabs that correspond to each of the surround channels when inserted into a Surround Bus (L/R | Ls/Rs | C | LFE | Surround Bridge).

In a standard 5.1 configuration:

1 L/R is the front Left and Right channels.

2 Ls/Rs is the Left and Right surround channels.

3 C is the Centre channel and LFE is the Low Frequency or Sub woofer channel.

The Surround Bridge tab is a panel that allows you to configure how each part is routed, linked and if it is enabled or not. This is useful for eliminating unwanted effects such as reverb on the LFE channel.

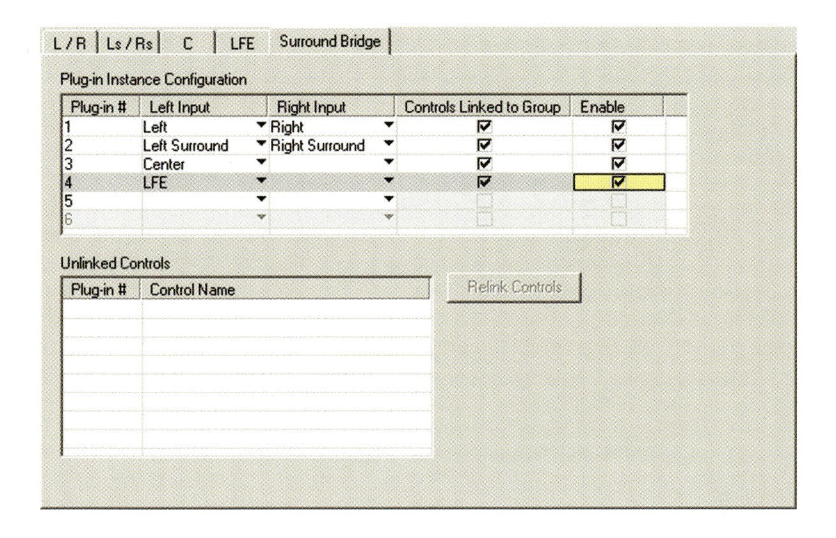

If you have more or less channels the tabs will be shown, routed and labeled accordingly.

Freezing Plug-ins

As soft synths and effects can use up valuable processing power, particularly on large Projects it's a good idea to render their Tracks as audio and turn off the synth or effect to release the resources back to the system for other duties. SONAR's Freeze functions can do just this quickly and easily.

To Freeze a Soft Synth

1. Use a Project with soft synths in then open the Synth Rack and click on a synth to select it .

2. Click the Freeze button and select Freeze Synth.

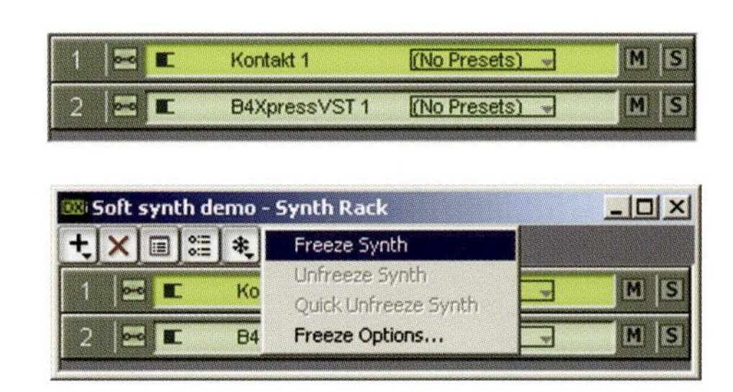

3 There is a list of options also available from this menu (Freeze Options) including the option to hide the MIDI Tracks when Freezing and also a noise gate type of tool which will cut out any silent passages in the Track thus conserving disc space.

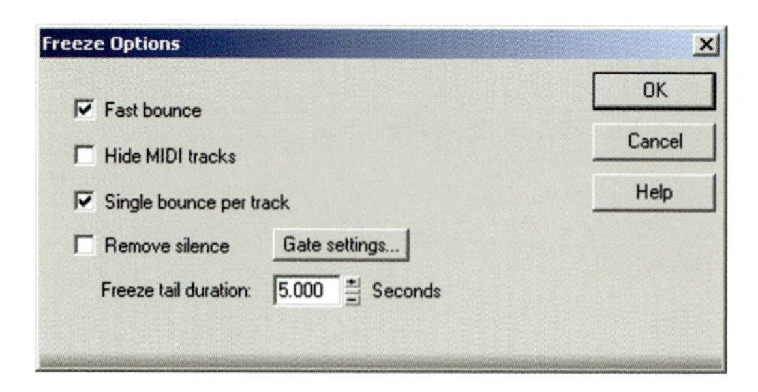

To Unfreeze a Soft Synth

1 Open the Synth Rack and click on a synth to select it.

2 Click the Freeze button and select Unfreeze Synth or Quick Unfreeze Synth, this quick option will store the bounced audio so that you can quickly Freeze the synth again whereas the Unfreeze option discards the audio. The Quick options are only available after the initial Freeze has taken place. Quick Unfreeze is useful if you want to compare any changes made between the original and the frozen audio as you can then use Quick Freeze without the Track having to go through the bounce process again.

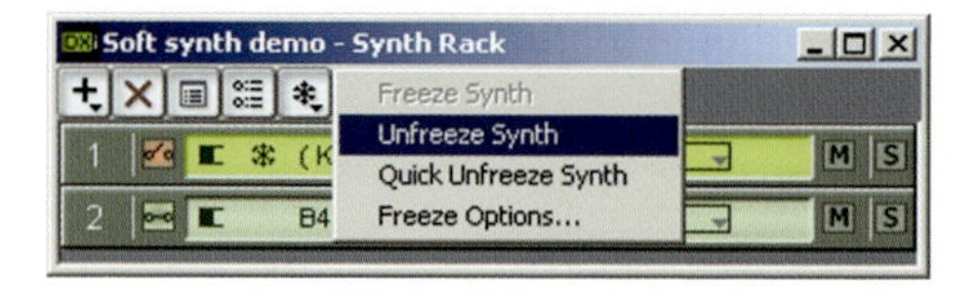

In the Options>Global page there is an option to Unload Synth on Disconnect which will release the system memory that the synth was using when connected that is not Frozen. Bear in mind that this will take a little while to reconnect when you Unfreeze the synth.

To Freeze an Effect

Tracks with effects inserted can also be Frozen to conserve resources, this is particularly useful for effects such as reverbs which use a lot of processing power.

Whenever possible it's better to share such effects by using them in a Bus but when used directly in Track Effect Bins here's how to Freeze them.

1 Select the Track to be Frozen by clicking on its Track number so that it's highlighted (1 Riff M S R).

2 Right-click on the number and select Freeze Track.

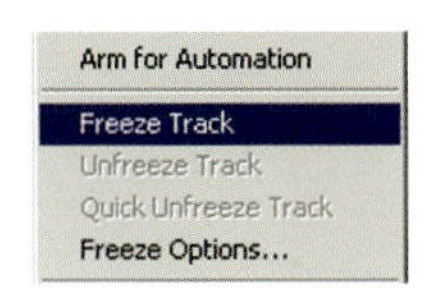

3 The Track will now be Frozen. Note that a stereo effect will result in a stereo Track when Freeze has been applied providing the stereo switch is turned on ( ).

To Unfreeze an Effect

1 Right-click on the Track number and select Unfreeze Track (Unfreeze Track).

2 The same Freeze Options and Quick Freeze/Unfreeze facilities are available from this menu as the Synth Rack Options.

Effects and Automation

Most plug-in effects can be Automated in real-time and by the use of Envelopes. Please see the chapter on Automation for more information.

CHAPTER 8

USING REWIRE

What is ReWire?

Most users of music technology will have heard of ReWire which is a system that allows two software applications to run together, in sync and for some controls to be shared by either application such as Play and Stop. ReWire was invented by a company called Propellerhead Software specifically to enable their ReBirth software, to work alongside other sequencers, being particularly useful to anyone wanting to use full audio tracks which ReBirth was unable to do. Other applications have since taken up the technology including Cakewalk's own Project5 sequencer.

ReWire treats one application as a Host and the other as a Client which in our case means that SONAR is the Host and the other application, Project5 or Reason for instance, would be the Client.

Once you have ReWired the two applications together you can switch between them quickly and work in both applications or route the ReWire Channels into SONAR (both audio and MIDI) and perform most functions from there. You can record audio and MIDI in SONAR and use all of its usual functions and features. Opening ReWire audio Tracks in SONAR and feeding the ReWire Client's instruments through them enables you to use the full complement of SONAR's mixing and effects system; if required you can Bounce any or all of the ReWire Tracks to audio Tracks in SONAR.

You can Freeze ReWire Clients in the same way as soft synths providing you have some information or data recorded in SONAR, if not then you could just insert a MIDI note at the end of your Project with a Volume of 1.

SONAR's implementation of ReWire is very elegant and easy to get to grips with as it uses the Synth Rack and makes the loading of a ReWire device as easy as loading soft synth. There exist similar options for ReWire devices as they may present multiple audio Outputs which the Synth Rack is already equipped to deal with from its Options dialog. Here's the basic procedure for loading a ReWire device.

1. Open an existing Project or start a new one and open the Synth Rack by clicking the DXi button in the Views Toolbar or selecting it from the View menu (DXi).

2. Click the Insert button (+) and select ReWire Device. This will only work if you have suitable devices installed otherwise it will remain 'grayed out'.

3. Select your chosen device from the menu that appears and the Options dialog will open.

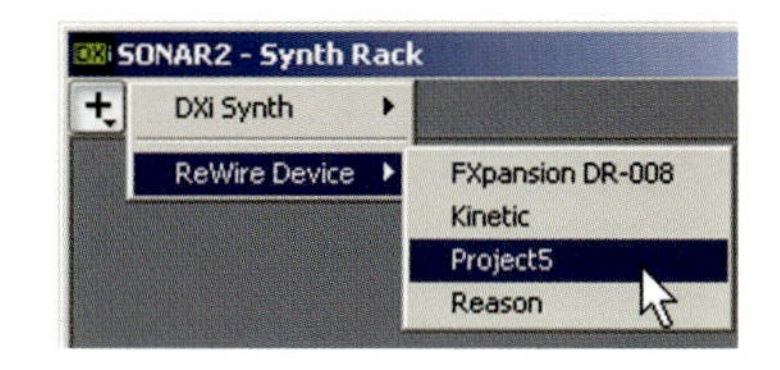

4. Make your choices here; if you want to play instruments in your ReWire Client from within SONAR then tick MIDI source Track.

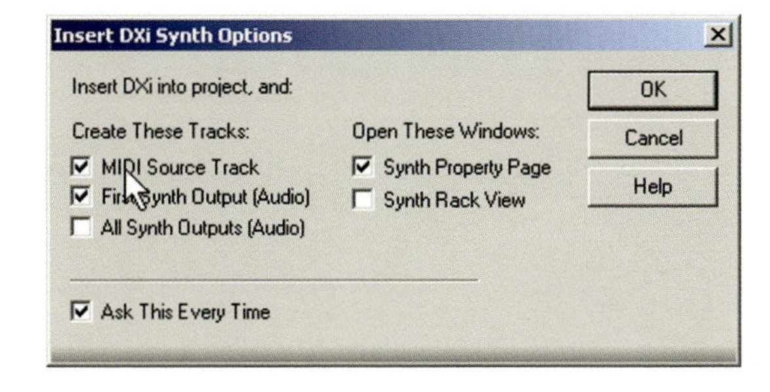

5. Be careful when selecting All Synth Outputs as you may find the ReWire Client opens 64 audio Tracks! If in doubt choose First Synth Output (Audio) as you can always add more later. If you need the full amount of audio Outputs then choose the All Synth Outputs (Audio) to load them. Note that you may need to route your ReWire instruments to the Audio Outputs in the ReWire Client itself as some applications will only route them to the first stereo pair.

6. Select Synth Property Page if you want to see the ReWire Client's interface then hit (OK).

7 The ReWire device will be started and automatically 'wired in' to SONAR.

8 With the ReWire Client's interface open (if it isn't open just double-click its name in the Synth Rack), load a project into it (there are usually demo tunes or tutorial files that you can use).

9 Minimize the ReWire Client's interface then click Play (▶) or press Spacebar in SONAR. The ReWired devices tune will play.

10 Now adjust the tempo in SONAR by clicking the Insert Tempo button ( ) in the Tempo Toolbar then typing in a new tempo and clicking OK. Both SONAR and the ReWire Client will now play at the new tempo.

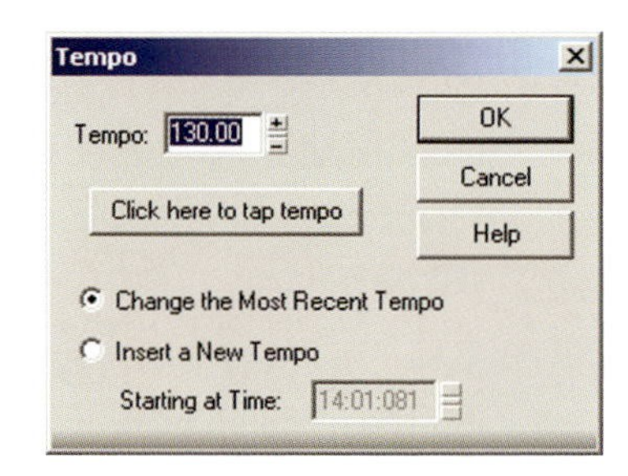

11 You can open the ReWire Client when you're working and use its Play and Stop controls as well as being able to tweak any of its parameters as if it was running in standalone mode. Record functions remain specific to their own application.

12 You may find that playback jumps to the Start when you hit Stop but you can set up looping points if you need to work in a specific area or alternatively start playback in SONAR and then click in its Time Ruler to move to that position. An even better workaround is to add a little data into the SONAR Project such as a MIDI note which will allow normal operation. Place the data at a late time in the Project (after the end time) and you'll be able to start from anytime before that point in SONAR.

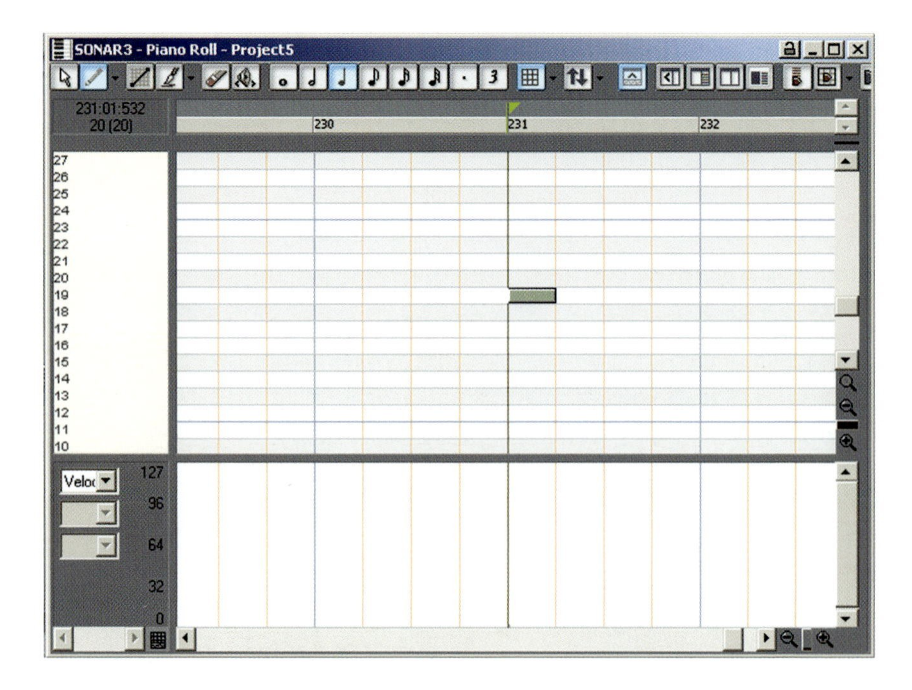

13 If you want to save your edits then you must save both parts (the ReWire project as well as the SONAR one) in their respective applications.

14 When you have finished, close the ReWire application first, then SONAR. If you don't then you'll see this reminder!

Playing your ReWire devices over MIDI can be achieved from SONAR as if you were using a soft synth either using the MIDI Track that was inserted with the ReWire device or adding another MIDI Track and setting it accordingly. Set SONAR and the ReWire device up first then . . .

1. Make sure that the Track's Output field is set to the ReWire devices Input, for example 3-Project5 (3-Project5).

2. Choose a MIDI Input Port and Channel or use Omni to use any incoming MIDI signal.

3. Routing to the ReWire instrument may vary from one application to another so here are a couple of examples:

Example 1

Project5 – From within Project5 right-click on a synths name and go to ReWire MIDI Input then select a MIDI Channel number. Do this for each instrument and select a different Channel for each one.

Minimize Project5 then use the Channel parameter in SONAR to route to the instrument on the corresponding Channel. Depending on the instrument you may also be able to change its Bank and Patch using the relevant fields.

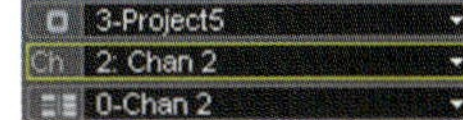

Reason – When you have Reason running over ReWire it will present its instruments by name in SONAR's Channel field so selecting them is easy.

Another useful feature of ReWire is that you can bring the audio Outputs from each device into SONAR on a separate audio Track. Here's a couple of examples of how to do it . . .

Example 2

Project5 – Simply add an audio Track to SONAR and from its Input field choose Project5 then a Project5 Synth number or device.

They appear as stereo ports and will configure stereo Tracks automatically. The Synth numbers correspond to the numbers shown in Project5 alongside the synth names (3 PSYN [7]).

You can now treat these Tracks as if they were ordinary audio outputs and add effects, Automation and mix as required.

Reason – Open Reason, hit your Tab key to reverse the rack then patch your chosen device into the Reason Hardware Interface either into a pair of channels for stereo or a single channel if it's a mono-device.

Back in SONAR insert a new audio Track or two to suit the device then click on the Input field and choose the corresponding Reason Channel number. If you have a stereo pair then repeat this for the other channel and don't forget to Pan them left and right to suit. You can now treat these Tracks as if they were ordinary audio outputs and add effects, Automation and mix as required.

Recording the ReWire Output

Providing you've worked your mix exclusively in SONAR and your ReWire Client you can perform a Bounce to render the mix as a stereo master. To do this . . .

1. Choose Edit>Select>All then drag in SONAR's Time Ruler for the length of your song. If SONAR's Tracks are longer than the ReWire Client's Tracks you don't need to drag in the Time Ruler.
2. Go to the Edit menu again and choose Bounce to Tracks.
3. Set the parameters you require in the dialog that appears and click OK.

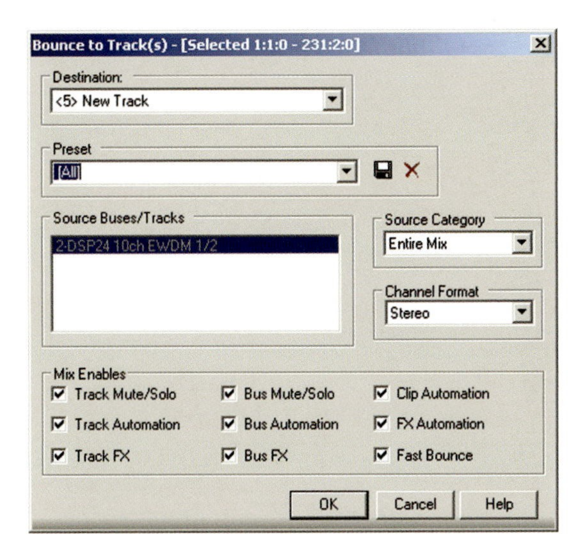

4 The Bounce will be performed and you'll then have a stereo audio Track of the whole lot, both SONAR and your ReWire Client's material.

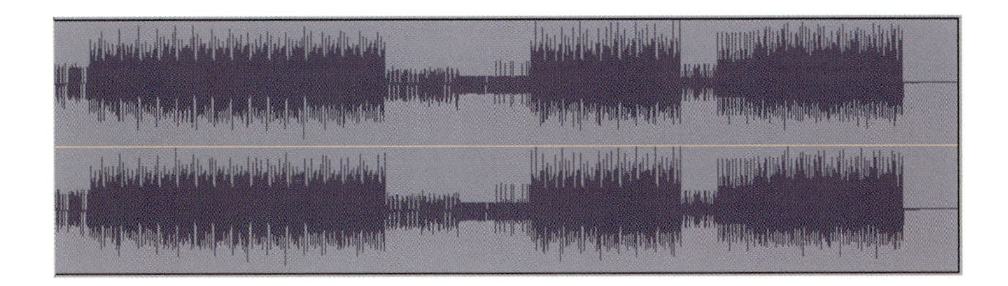

If you've routed anything to external hardware then you'll have to set up a stereo audio Track and record the sound in real-time or record individual instruments back in then perform a Bounce with them included.

CHAPTER 9
THE CONSOLE VIEW AND ROUTING

If you are used to using a mixing board then you may find the SONAR Console suits your way of working quite easily. If you're not used to using a mixer then it's worth spending a little time getting acquainted with the Console as it provides a different overview of your Project than the Track View which is often preferable when nearing completion. I find the main reason to use the Console View is that I can see many parameters at once that are often hidden in the Track View; the FX Bins for instance are much easier to see in the Console's vertical Track Strips. Strips may be re-ordered by dragging to a new location to make viewing even clearer.

The Console View is opened by clicking its button in the Views Toolbar ( ) pressing Alt+3 or by selecting it from the View menu. It is user configurable from the panel on its left side which contains 11 buttons for toggling the various features as follows:

1. Allows you to widen or narrow all Track Strips ( ).
2. Shows or hides all meters. It also has an Options dialog for configuring your meters obtainable by clicking the arrow next to it ( ).
3. Shows or hides the Input controls which includes Trims on Channel Strips and input Gains and Pans on Buses ( ).
4. Shows or hides the EQ Plot ( ).
5. Toggles between the three EQ control states On, Off and an extended set of parameters ( ).
6. Toggles between FX Bins on, off and on with assignable FX controls (FX).
7. Toggles between the three Send control states Off, two sets of Send controls visible, four sets of Send controls visible ( ).
8. Shows or hides the Mute, Solo, Record Arm, Phase, Interleave and Input Echo controls (MSR).
9. Shows or hides the Pan controls for Tracks and Buses ( ).
10. Shows or hides the Volume faders and their meters ( ).
11. Shows or hides the Output routing fields (O).

The Console is split into three areas that can be resized by dragging the vertical splitter bars. On the left is the Channel Strips section, in the middle is the Buses section and on the right is the Main Output section. Both audio and MIDI Channels will appear in the Channel Strips in the same order as they are in the

Track View (but Track Folders are not available in the Console View).

Press M to invoke the Track Manager where you can select which Strips are visible from within the Console View by ticking and unticking the relevant boxes; the Toggle controls allow you to select groups quickly. To Hide individual Strips quickly just right-click on the Strip and select Hide Track.

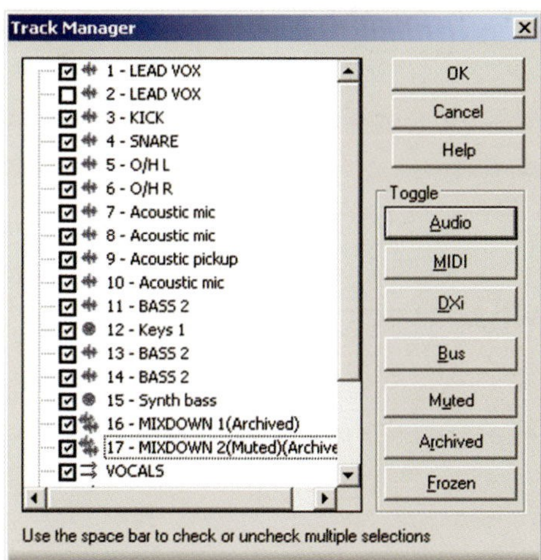

Main faders can be locked together as stereo pairs by clicking the lock button ( ) at the bottom of their respective Strips.

All controls are easy to operate by clicking and dragging, and can be reset to their default positions by double-clicking on them. Some controls such as Sends, EQ's and EQ Bands may have to be activated by clicking buttons before they will have any effect ( ).

FX Bins work in the same way as they do in the Track View and plug-ins are inserted by right-clicking and selecting the effect from the available ones in the Audio Effects menu. They can be dragged up or down into a different order if required and turned on or off by clicking their switches (PSP_MixPress) and edited by double-clicking their names to open the relevant plug-in.

New Tracks can be inserted by right-clicking on any Strip and choosing an audio or MIDI Track from the menu.

Automation can be performed in this View by recording movements or taking Snapshots (see Chapter 10, Automation).

Audio Track Strips

Audio Track Strips are very similar to their hardware counterparts and in general you can imagine the signal coming into the top of the Strip and traveling down through it to emerge at the bottom then follow the routing shown in the Output To field. This isn't strictly how it works as some controls such as Sends may be routed Pre or Post fader but it is a good simple generalization for now. The name of the Strip can be changed by double-clicking in the name field at the bottom of the Strip and typing a new name in.

MIDI Track Strips

MIDI Track Strips contain all of the controls that are available from the Track View but in Strip form. There isn't really a hardware counterpart for this but it does make for a very ergonomic way of working alongside the audio Strips. All controls work as expected and MIDI FX Bins allow insertion of MIDI effects plug-ins. These do not have activation buttons like audio effects but can be edited by double-clicking on their names to open the relevant plug-in. The Output routing may be changed by selecting another option from the Output To field and the name of the Strip can be changed by double-clicking in the name field at the very bottom of the Strip and typing a new name in.

Track Sends

Track Sends can be used to take some or all of the signal from any audio Track and send it to another destination, usually a Bus or Main Output. This can be to apply an effect to several or indeed all Tracks as if it were an auxiliary Bus or maybe create a separate monitor mix that will be sent to a different Main Output or even to a Subgroup. The uses are many and the system is very flexible particularly because you can use the Send as Pre or Post fade.

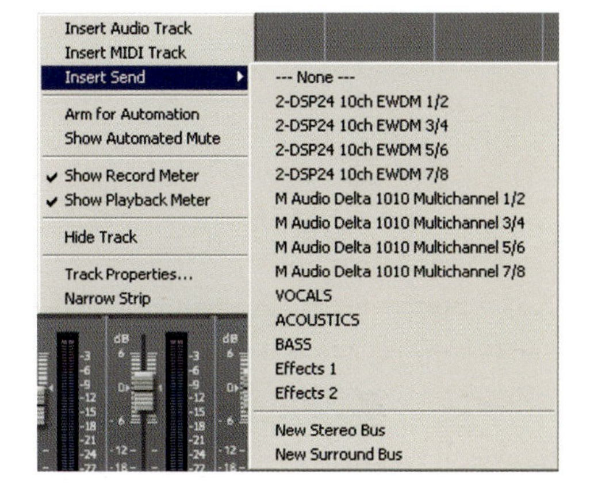

Track Sends can be inserted on any audio Track from the Console View (or the Track View) by right-clicking on the Track Strip (or its panel in the Track View) and choosing Insert Send. The menu that appears will show you all available options for the Send's destination including the option to add a New Bus.

The Send will be inserted (including a New Bus if chosen) but will be inactive until switched on using its button ( ).

The Send will show its destination which may be altered by clicking the adjacent arrow and by choosing Reassign Send to select a new destination or Delete Send to remove it altogether (note that to avoid feedback a Track or Bus can never be sent to the same destination as its direct Output).

Clicking the Pre/Post switch will change the Send's position in the signal chain to Pre fader (so the fader setting will not affect it) or Post fader (so the fader setting will affect it) ( ).

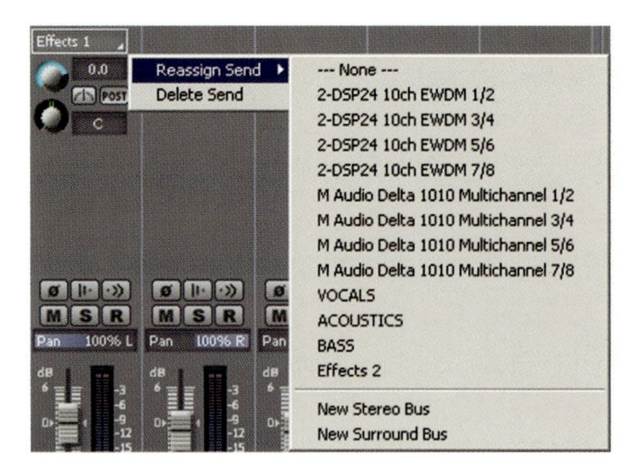

The Send Level can be adjusted using the Send knob and its Pan can be adjusted using the Pan knob. Both have visual indication of their current state shown as a colored ring, blue for the Level and green for Pan.

Buses

Buses can be used to route Tracks or other Buses into and in turn the Bus may be routed into another Bus or Main Output. Buses can also contain Sends which can be routed into Buses or Main Outputs. Obviously this provides a very flexible system enabling extremely complex routings to be created but it also makes it easy to create simple routings that are commonly required such as Subgroups and Aux Buses for effects.

Buses can be inserted from the right-click menu either in the center section of the Console View or the control panel of the lower Bus Section of the Track View. A New Bus can also be added when a new Send is inserted by choosing New Bus from the destination options.

Double-click in the name field and type in a suitable name.

Inputs to Buses are set by selecting the Bus name from the Track or Bus Output fields of any Tracks or Buses that you want to route into it. Note that this will send the signal entirely into the Bus and any previous routing will not be present so if you had a Track routed to a Main Output then changed it to a Bus it may not appear at the same Output as it will now go to the Output that the Bus is routed to.

Bus Outputs are set by choosing a destination from the Output To field at the bottom of each Bus Strip. This can be a Main Output or another Bus.

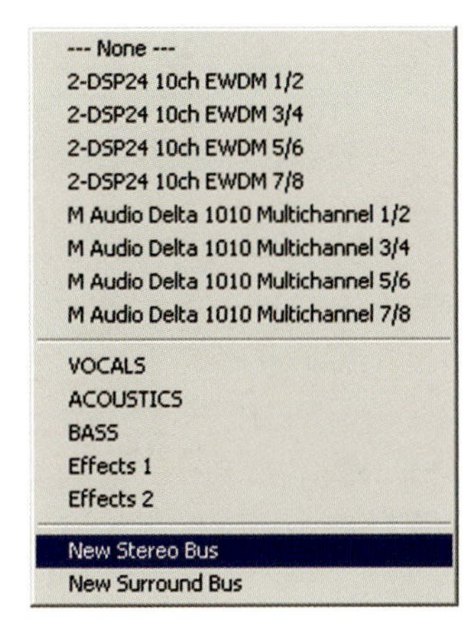

In this example the top section shows all of my available soundcard outputs then below this are the Buses that I have already set up namely: VOCALS, ACOUSTICS and BASS Subgroups and Effects 1 and 2 which are used as Aux Buses. The options for New Stereo Bus or New Surround Bus are also available at the bottom.

Mains

The number and type of Main Outputs will depend on which Outputs are available to SONAR. You can define these by highlighting them in the dialog at Options>Audio>Drivers. You may have to restart SONAR to make the changes take effect. You cannot insert any effects or Sends into Main Outputs, use Buses to do this. Mains are used for basic control over Level and Pan of the Output in question.

Creating an Aux Bus to Share an Effect

Use a Project with several audio Tracks for this then . . .

1. In the center section of the Console View, right-click and choose Insert Stereo Bus.
2. Right-click in the FX Bin of the Bus and insert an effect, a reverb is a good choice.
3. At the bottom of this Strip double-click on the Bus name and type in a new name such as Reverb 1 (Reverb 1).

4 Go to the left side of the Console View and right-click on an audio Track Strip, choose Insert Send then select the name of your new Bus (Reverb 1).

5 Click the ( ) button to activate the Send.

6 The Send Level will be at 0 dB so if the song is played it will be sending a high Level of signal to the Aux Bus. You can adjust this easily by clicking on the Send knob and dragging up or down.

7 If you want to, you can click the Bus Output To field and assign it to a different Main Output or Bus.

8 You can add Sends to any audio Track or Bus and route it to your new Aux Bus either single Tracks or maybe a whole Subgroup, all with their own Levels.

Reverb can make a sound appear to be close or far away depending on how much is applied to the sound so use a lower Level for sounds that should appear close and add more for sounds that should appear farther away. You can also use EQ on the Aux Bus to stop the bass end becoming muddy and to enhance the higher range a little if you want to.

Creating a Subgroup

1 You need a Project with several Tracks for this, preferably ones that lend themselves to a Subgroup such as drums or an ensemble.

2 Open the Console View and in the center section right-click and choose Insert Stereo Bus.

3 Double-click in the name field at the bottom and type Subgroup 1.

4 In the left-hand section choose the first Track to be part of the Subgroup and click its Output To field; change it to Subgroup 1.

5 Continue to do this for each Track that is going to be part of the Subgroup.

6 Set the Subgroup 1 Output To field to a suitable Main Output;

if you have several, then one that isn't being used would be good.

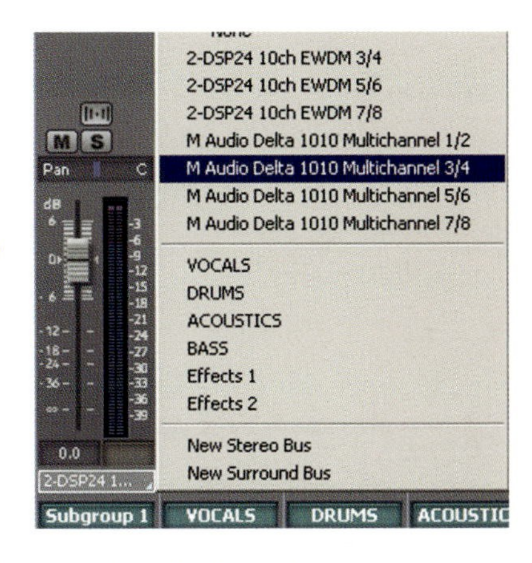

7 Play the Track and you can now set the Level, add effects and change all the parameters for the Subgroup using just the single Subgroup 1 Track Strip.

8 If you had a Bus with an overall effect in, such as a reverb that was going to be used on all Tracks, you could add a Send to the Subgroup, route it to the effect Bus and add whatever Level of reverb you wanted.

Surround Options

SONAR has extensive options for mixing in Surround Sound with profiles using any number of channels up to 8.1 available from the Surround tab of the Options>Project dialog.

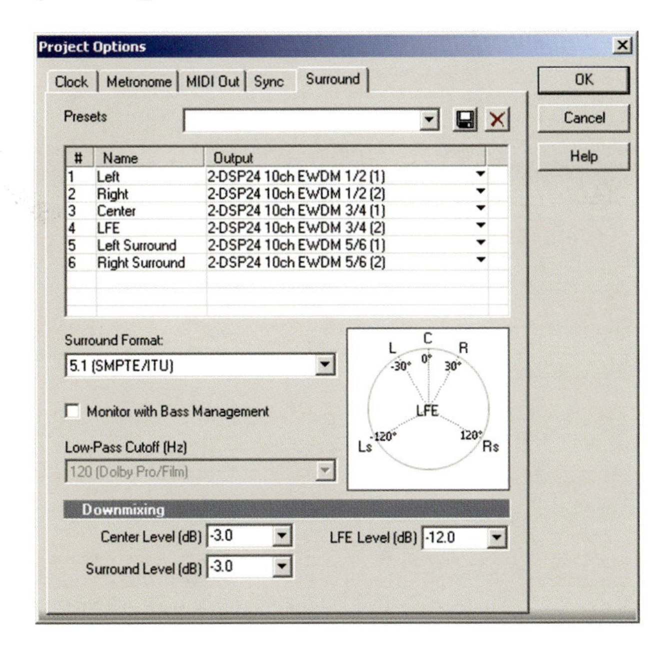

Here you can define the important parameters necessary for effective mixing in a surround format but bear in mind that if you want to hear your efforts on another system such as a DVD player then you'll have to use an encoder which you need to purchase separately. SONAR has a trial version of the excellent diskWelder BRONZE package included on the install CD which will allow you to create DVD audio.

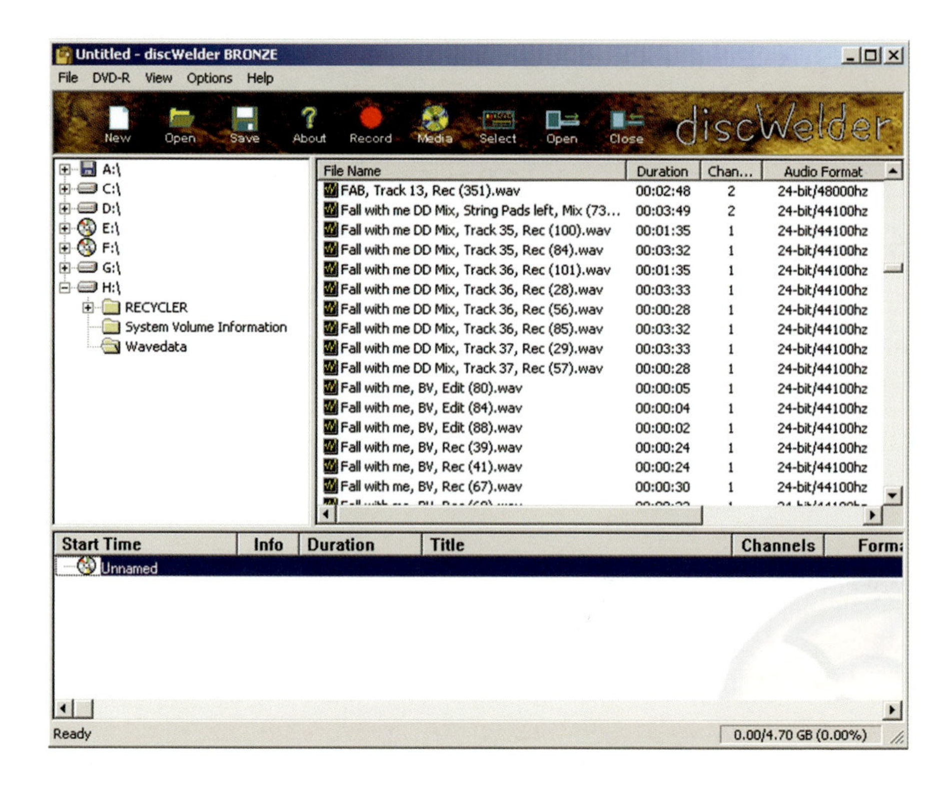

If you need serious DVD surround authoring (for commercial film, etc.) then you may want to look at the high-end packages such as SurCode.

Information about the full range is available from www.MinnetonkaAudio.com

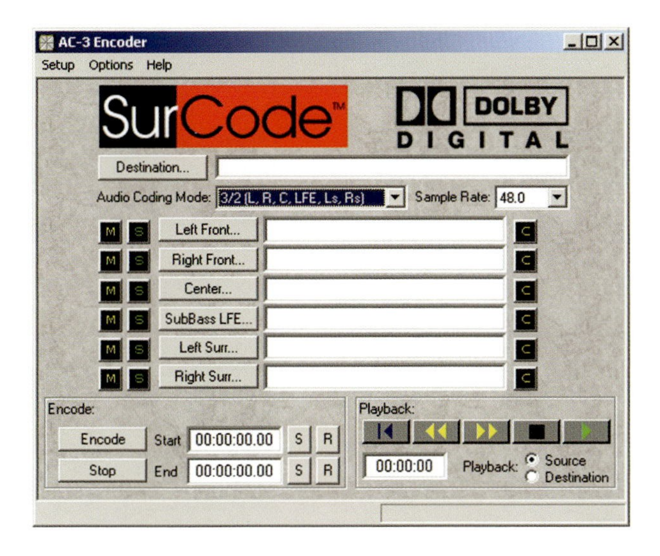

Surround mixing is available using Surround Buses which can be inserted in the same way as Stereo Buses.

Sends can be inserted into audio channels which can in turn be routed to the Surround Bus.

Audio channels may be routed directly into a Surround Bus by choosing the Bus from the channel's Output field.

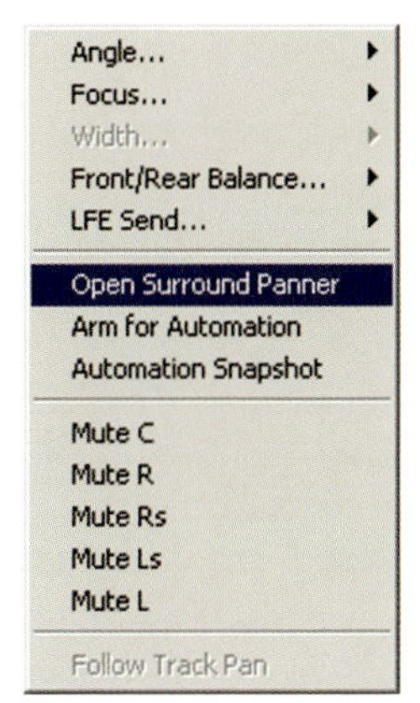

The sound from a Send or channel appears in the Surround Panner which appears a little on the small side when viewed in a channel strip ( ) so to work effectively we need to open the full view of it. This is achieved by right-clicking on the small version and selecting Open Surround Panner from the menu that appears.

You'll also notice that this menu includes many options for advanced surround mixing.

These options are also available by right-clicking on the full-size Surround Panner.

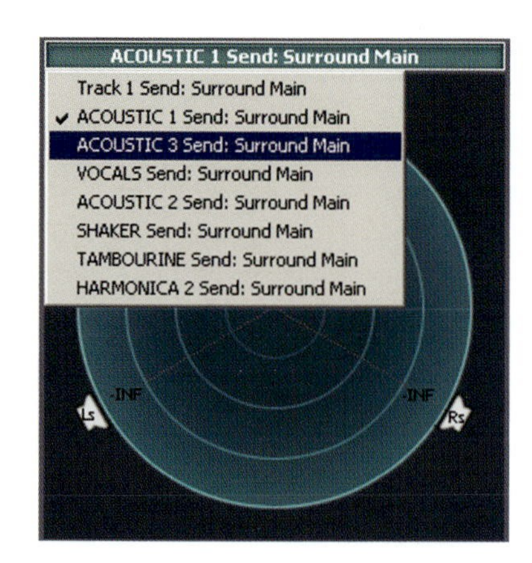

When open you can edit any channel coming into the Surround Panner by selecting it from the Outputs menu along the top of the Panner.

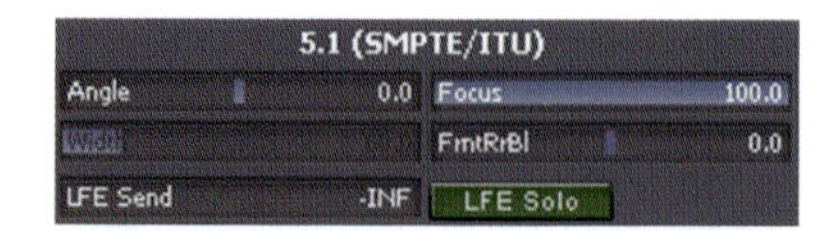

The lower control panel consists of sliders which can be dragged to adjust Angle, Focus, Width, Front to Rear Balance and LFE Send level. They can also be remote controlled using the Remote Control dialog available by right-clicking on them.

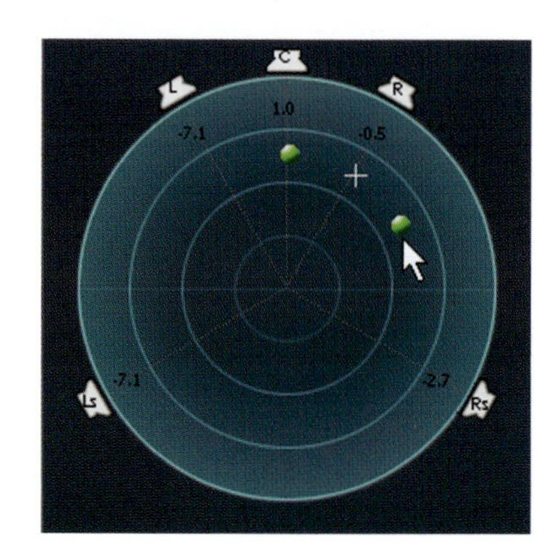

The main body of the Panner is a graphic representation of the surround area with a pair of spheres (or a single sphere for mono signals) and a crosshair to show the current position of the signal.

Dragging anywhere within this area will move the spheres and the signal from the currently selected channel around the surround field, if your hardware is set up correctly you'll hear the sound moving as you drag.

Clicking on a speaker icon will toggle its state. A single-click will mute or unmute it (turns dark gray) while a double-click will solo or unsolo it (turns green) ( ).

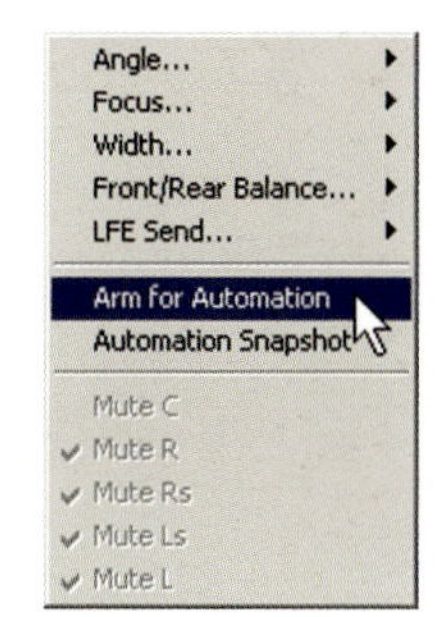

If you have a joystick attached and configured correctly (see online Help) then you can use this to control the Panner.

The Surround Panner controls can be Automated by right-clicking on the Panner and selecting Arm for Automation then carrying out the normal procedures for recording Automation.

Grouping Controls

It's possible to create control Groups for many parameters in SONAR either in the Console View or the Track View. The principle is the same in either View so I'll show you how to do it in the Console View while we're on the subject!

Creating a Fader Group

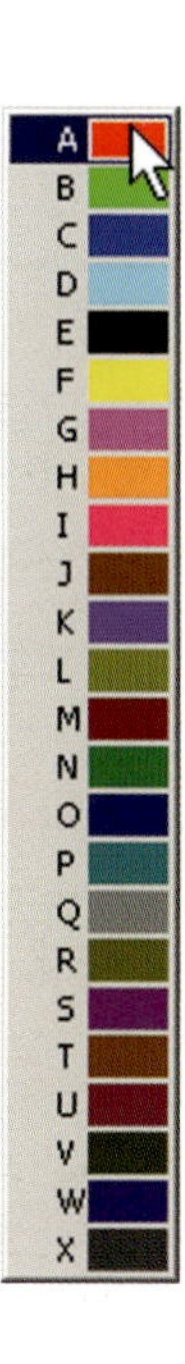

1. Open a Project with a few Tracks up and running then fire up the Console View.
2. Right-click on a fader and select Group, then a number/color combination such as 'A'; you'll see a small square of that color appear at the bottom of the fader.
3. Now choose another fader and do the same, choosing the same Group as before.

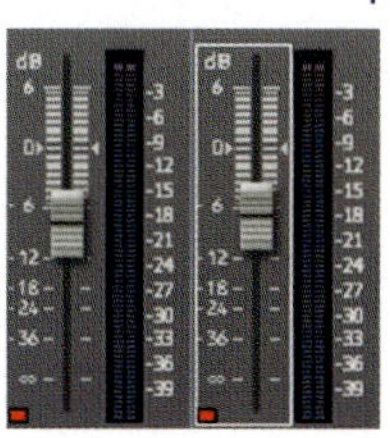

4. Now when you drag either fader in the Group the other one will follow, cool!
5. To further enhance your control, right-click on a fader in the Group and choose Group Properties.

6 You can choose three states for the faders: either Absolute, Relative or Custom. You're already using Absolute so try Relative then come back to this dialog and select Custom.

7 With the first fader's name highlighted set its Start Value to 0 and its End Value to 127.

8 Highlight the second fader and set its Start Value to 127 and its end value to 0.

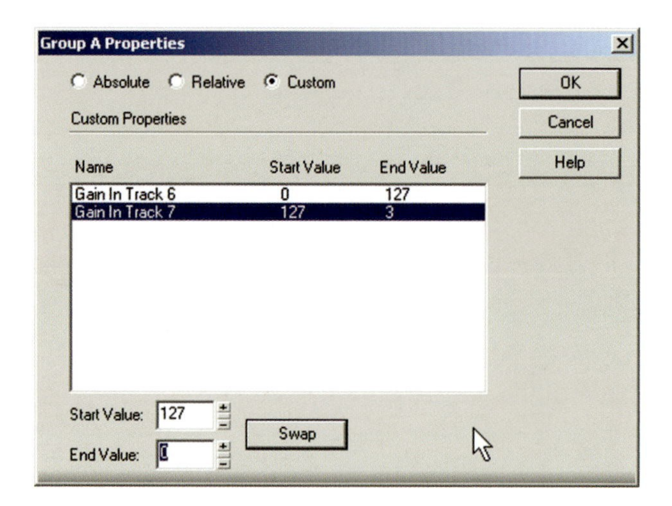

9 Click OK then drag one of the faders, you've now got a pair that crossfade by moving either one up or down, easy isn't it?

10 To Ungroup a control just right-click on the control and select Ungroup from the menu.

Pan, Mute or Solo Groups can also be created and customized in the same way speeding up many operations.

Remote Control

If you have a MIDI controller attached to your system then you should be able to control many parameters remotely with it; even a data entry slider or a

Modulation wheel can be used for this. A dedicated Control Surface system is available within SONAR but there's also a quick way of bonding a hardware control to a software control using this method . . .

1. Ensure your controller is sending data correctly, you should see the MIDI Activity Meter lights in the taskbar working when the controller is moved ( ).

2. Right-click on a control (most will be available) and select Remote Control from the menu.

3. You can enter the hardware controller parameters if you want but easier still is to click the Learn button (Learn).

4. Move the hardware control and you'll see SONAR detect its signal as the parameters update in the Remote Control dialog.

5. Click OK. You can now move the hardware control and the software control will follow its movements. If the control is part of a Group then the other members of the Group will also move with it.

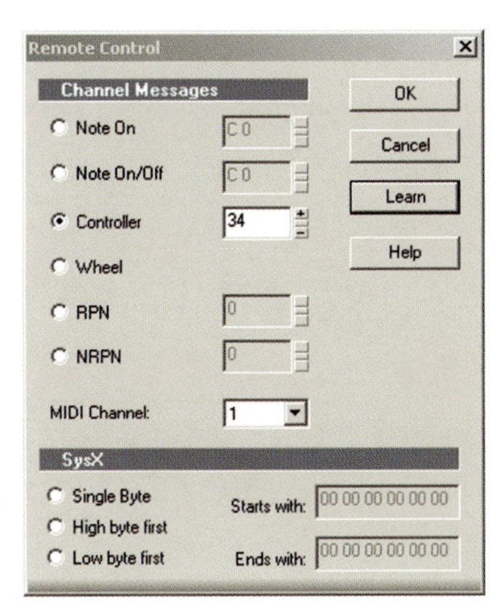

CHAPTER 10
AUTOMATION

Not so many years ago all mixing and effects adjustments had to be performed completely manually, if you wanted a fade in then you had to move the faders upwards at the correct time, to the right position and end up there at the right time. This is a very simple move when compared to the full amount of operations that might be required for a large scale project such as a movie score or even a large multitrack of a band where several operators may be working to a set of cues and following the time readout from a tape machine. Recent advances have brought us a high level of automation for hardware such as mixers but it simply cannot compare with the total flexibility and editability of software Automation.

SONAR has a fantastic level of automatable parameters ranging from the basic mix parameters (Volume, Pan and Send controls) to specific effect controls and soft synths. The actual Automation data is represented by visible Envelopes (although they can be hidden), which may be edited to a high degree, copied to other Tracks and even reassigned to different parameters. They can also be used on Buses enabling a further depth of control over your mixes. Both audio and MIDI parameters can be Automated which means that you can use Automation for soft synth Input and Output Tracks as well, an example of this would be to Automate the Volume of a MIDI Input Track to a soft synth so that you didn't affect the Volume of any other Tracks using the same synth which would occur if you Automated the audio Output Track of the synth.

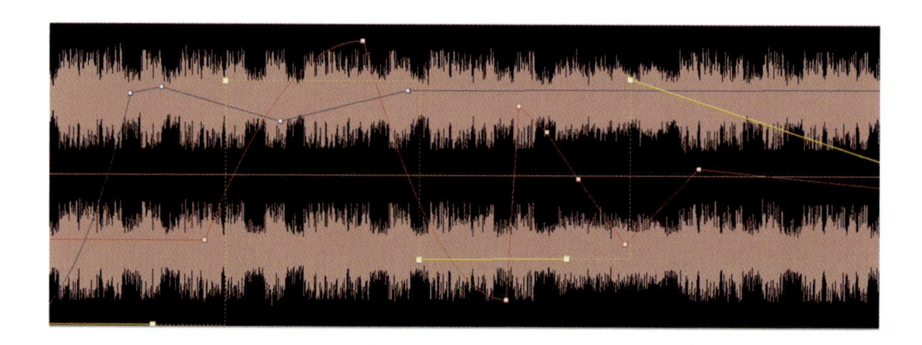

There are two methods of creating Automation; Recording it and Drawing it in. Either method results in an editable Envelope but sometimes recording may be preferable due to its more tactile approach and the fact that you can hear what's happening as you do it. At other times you may require the finer details available when drawing directly onto the Clip using the visible wave as a guide although this can be done after recording to edit your Automation.

A special Envelope tool is available which is very useful when editing as it's quite easy to accidentally perform a function you didn't intend to do when working with Envelopes if you don't quite click on the line: This can become quite infuriating if you keep moving Clips around inadvertently. The tool can be selected by pressing the E key or clicking its icon ( ) in the main tool selector at the top of the Tracks Pane.

The arrow next to the button provides options for showing and hiding groups of Envelopes, useful if you have a lot on a single Track.

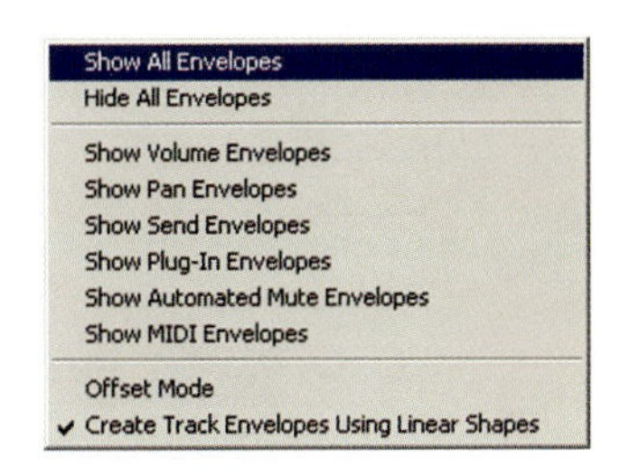

We'll have a look at both methods starting with recording as drawing is also the same method that you will use when editing any recorded data so it's logical to put it second.

If it isn't already visible then open up the Automation Toolbar from the View>Toolbars . . . menu ( ).

This will make it easy to disable any Armed controls by a single-click on the Disarm All Automation Controls button ( ).

Playback of Automation can be Enabled or Disabled by clicking this ( ) button. The other buttons will be explained later in this chapter.

Recording Automation on an Audio Track

1. You'll need a Project with some audio running for this so load up a suitable Project.

2. Select an audio Track to perform some Automation on and open its control panel on the Tracks Pane. You can Solo it if you like.

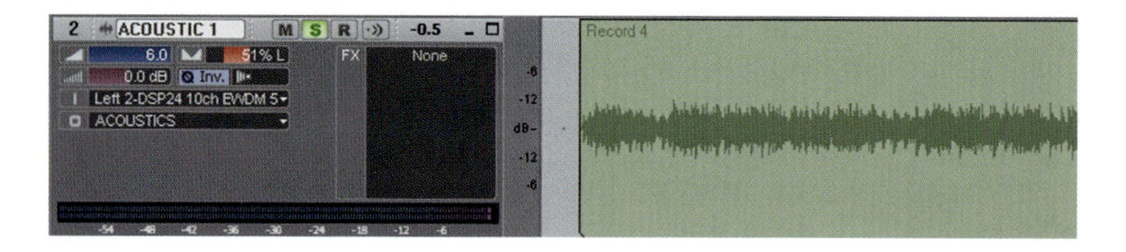

3 To select a single control (such as Volume) right-click on the control and choose Arm for Automation from the menu.

4 To select all Automatable controls click on the Track's number (2) to highlight it then from the Track menu (or the right-click menu) choose Arm for Automation.

5 Your chosen controls will now have red surrounds to show that they are active and a new Record Automation button appears in the Transport controls ( ).

6 Click this button then move your controls as required, hit Stop or Spacebar when you've finished.

7 The Automation Envelope(s) will appear superimposed on your audio Clips. If you have moved more than one control then there will be a different colored Envelope for each control.

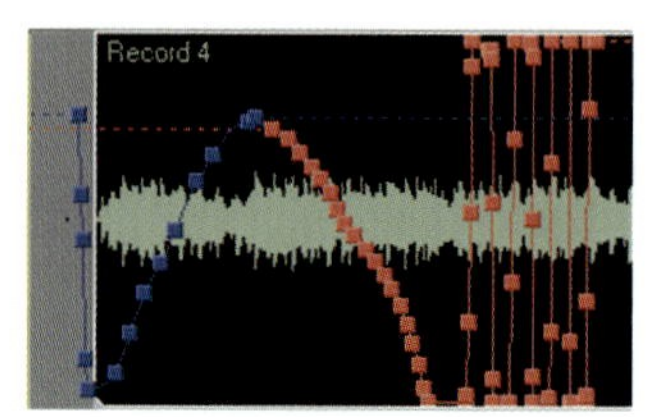

8 You can now Disarm all controls by clicking the Disarm button in the Automation Toolbar or unticking Arm for Automation in the Track or right-click menu ( ).

Recording Automation on a MIDI Track is exactly the same except the controls available are obviously a little different.

To see exactly where your Envelope is lying just hold the cursor over it for a second or two and you'll see a readout appear that is specific to the Envelope, for example, %L or %R for Pan and dB for Volume.

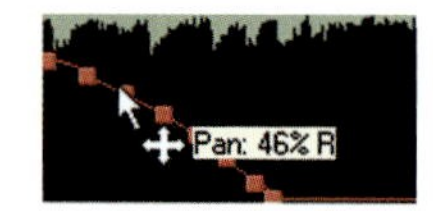

Drawing Automation

Drawing Automation can be performed on either MIDI or audio Tracks, as with recording Automation only the controls are different. I'd recommend using the Envelope tool ( ) when doing this as it's very easy to accidentally move something else if you don't. Here's how to draw a Pan Envelope on either MIDI or audio Tracks.

1 Load up your Project then expand a Track so that it's easy to see its Clips.

2 Right-click on the Track in the Clips Pane (it doesn't matter if you're over a Clip or not for this) and select Envelopes>Create Track Envelope>Pan from the menus that appear.

3 You'll see a thin orange line appear running horizontally through the Track, its position will depend on your current Pan setting; above the center line is left and below it is right.

4 To adjust the Envelope we need to insert Nodes that can be dragged around, to do this double-click on the Envelope and a Node will appear.

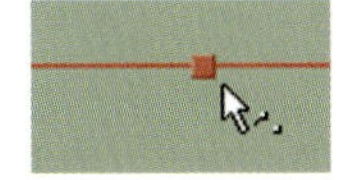

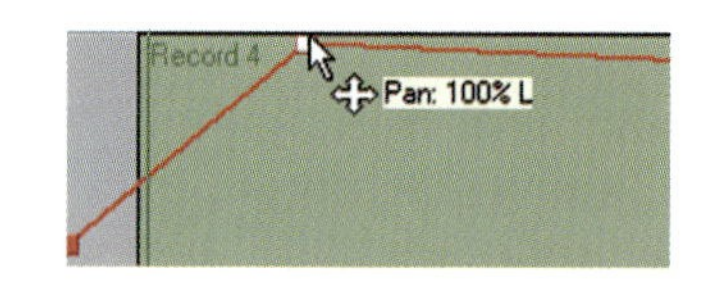

5 You can now click on the Node and drag it to move the Envelope in any direction. Drag it to the top to Pan the Track fully left.

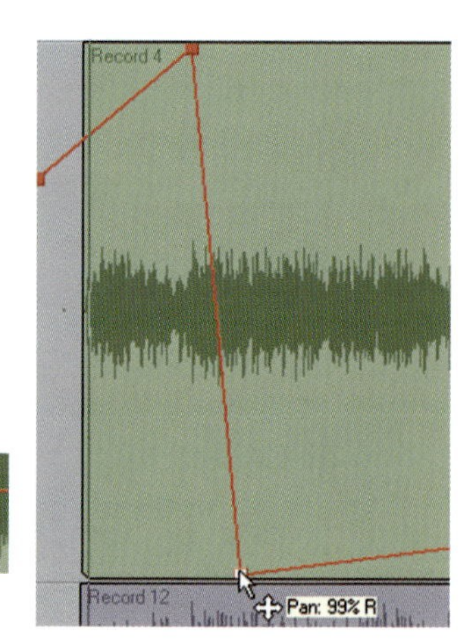

6 Insert another Node close to the first one and drag it down to the bottom of the Clip about a bar further on than the last Node.

7 Now add another Node a bar further on then double-click it. This will send it to its default position in the Track's center.

8 Now you can play the Project and hear the Track Pan from left to right and then back to the center as it follows your Envelope. You'll also see the Pan slider move as it updates to follow the Automation Envelope ().

You can also right-click on a single Clip and choose Envelopes>Clip to apply a simple Gain or Pan Envelope to a single Clip.

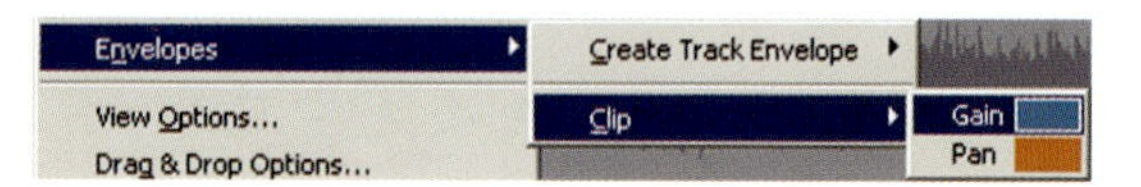

Automating Surround Parameters

If you have a Surround Bus running in a Project then you can Automate the parameters of the Bus in the same way as any other Track or Bus; right-click on the Track Strip in the Console or Track View and choose Arm for Automation to enable recording or alternatively right-click in the right-hand section of the Buses Pane in the Track View and select an Envelope to insert and edit.

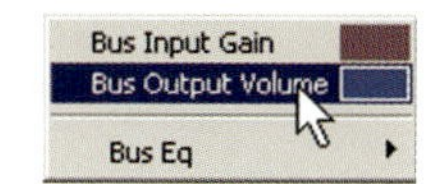

Surround Panner settings are a little easier to Automate if you follow this procedure:

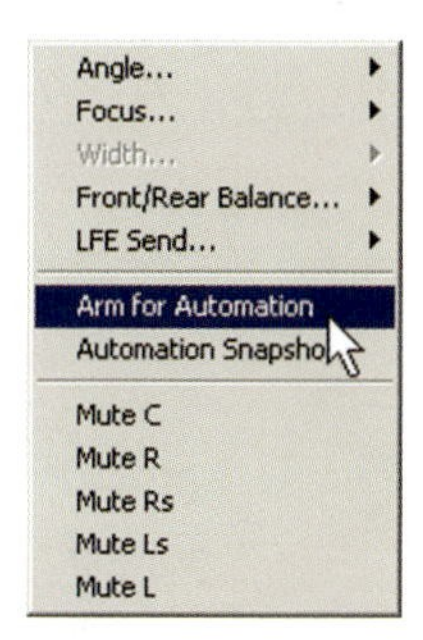

1. Choose a Track with a Surround Panner Send inserted and right-click on the Panner ( ).

2. Select Open Surround Panner to invoke the large scale Panner which will then appear.

3. Right-click on the Panner and choose Arm for Automation.

4. Click the Record Automation button ( ) then drag the Panner controls as required.

5. Hit the stop button or Spacebar to end recording.

6. Right-click on the Panner again and deselect Arm for Automation.

7. You can now edit the Automation Envelopes if required.

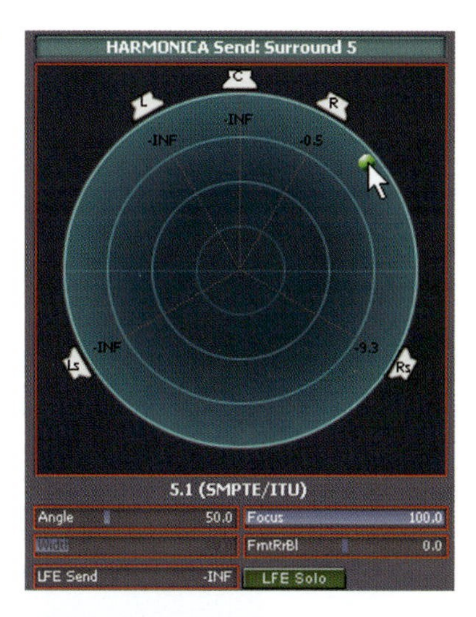

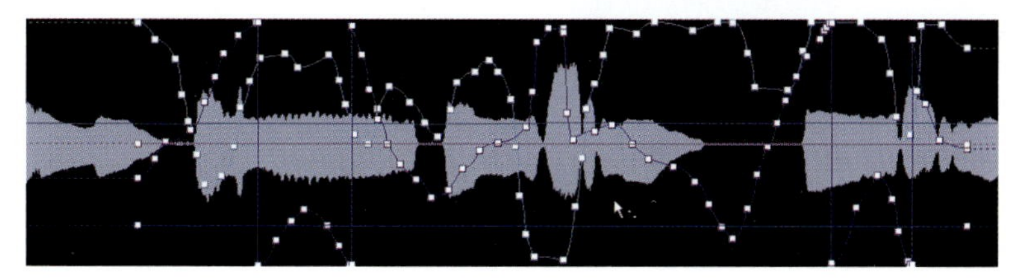

If you have a joystick controller attached and configured then you can use it to perform the Automation recording if you prefer.

Snapshot Automation

Another variety of Automation is the Snapshot which can make the setting of several parameter changes at a given point in time, easy. You need to Arm all controls that will be adjusted; the quickest way to do this is to use Edit>Select>All then Track>Arm for Automation to Arm all available Track controls.

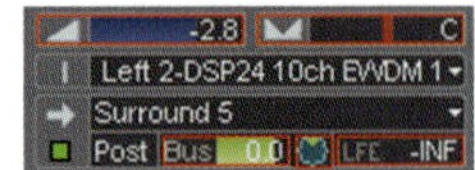

1. Click anywhere in the Clips Pane that you want to make a change to move the Song Position Pointer to that point. You may have to adjust the Snap To setting if it's incorrect.

2. Adjust the control parameters as desired.

3. Click the Automation Snapshot button and you'll see new Envelopes and Nodes appear at the point you took the Snapshot.

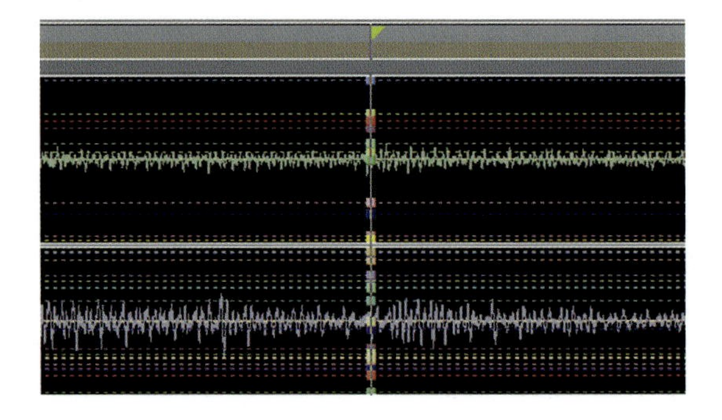

4 Move to the next position, make your new adjustments and click the Automation Snapshot button.

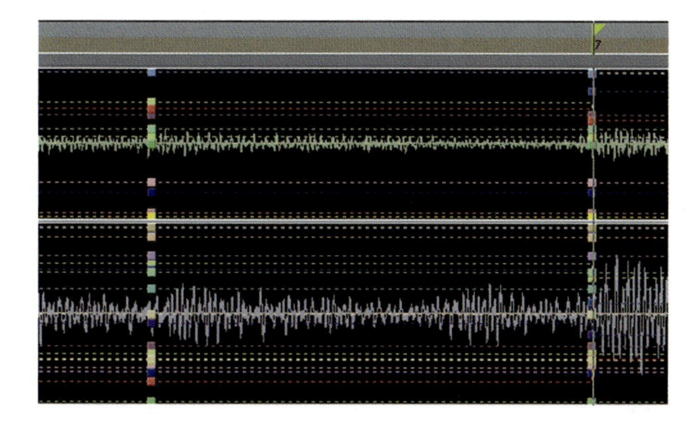

5 When you've finished you can Disarm all of the Armed controls by deselecting Arm for Automation in the Track menu.

You can also take specific Snapshots by right-clicking on a control and selecting Automation Snapshot. You don't need to Arm the control to do this so it can be much quicker if you've just got one or two changes to make.

Automating Effects Plug-ins

Most effect plug-ins can be automated which can be very useful for switching them in and out at specific points or changing their Levels or other parameters. They can be Automated using recording or by drawing in and editing Envelopes. To Arm a plug-in for recording Automation:

1 Insert the plug-in effect to an FX Bin.

2 Set up its controls to work with your Track.

3 Right-click on its name in the FX Bin and select Arm Parameter; if the option is grayed out it means that the plug-in is not automatable.

4 Tick the boxes for any parameters that you want to tweak or click Arm All to make them all ready then click OK.

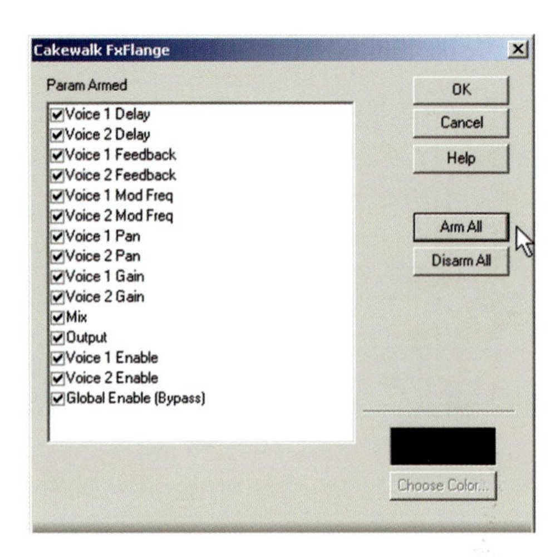

5 Make sure the effect is open and visible then click the Record Automation button in the Transport Toolbar ( ).

6 Adjust your effect as required as the song plays to record the Automation.

7 When you've finished click the Disarm All Automation Controls button in the Automation Toolbar ( ).

If you want to draw Envelopes instead of recording then:

1 Insert the effect into the FX Bin.

2 Right-click in the Clips Pane and choose Envelopes>Create Track Envelope then select the plug-ins name.

3 A dialog box will open enabling you to tick boxes to select any control Envelopes that you want to create, do this then click OK.

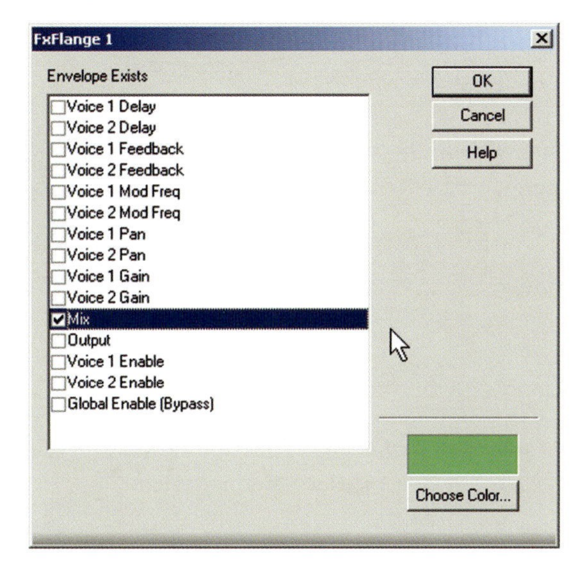

4 Your chosen Envelopes will now be visible on the Track ready for editing.

MIDI Envelopes for Soft Synths

It's possible to control soft synths or external MIDI hardware by using MIDI Envelopes to adjust the device's parameters. The parameters available will depend on the Output or soft synth itself and may be named to match the labeling on the synth or just standard MIDI parameter names or numbers.

To create a MIDI Envelope:

1 Select your MIDI Track in the Track View and right-click in the Clips Pane of the Track.

2 Choose Envelopes>Create Track Envelope>MIDI. . . .

3 A MIDI Envelope dialog opens where you can select the Type of controller, the Value and Channel. Do this then click OK.

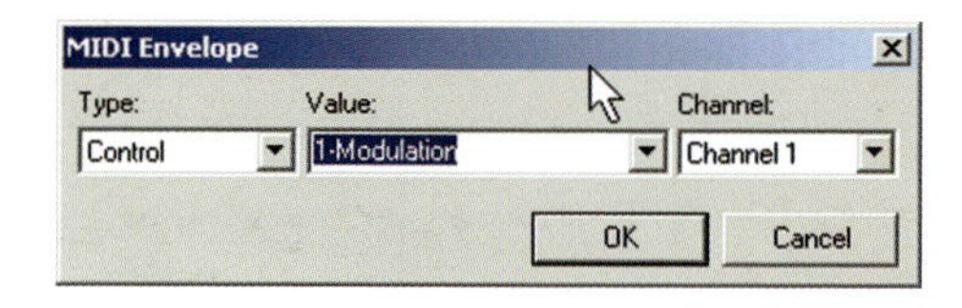

4 An Envelope will appear corresponding to the selected controller which can now be edited.

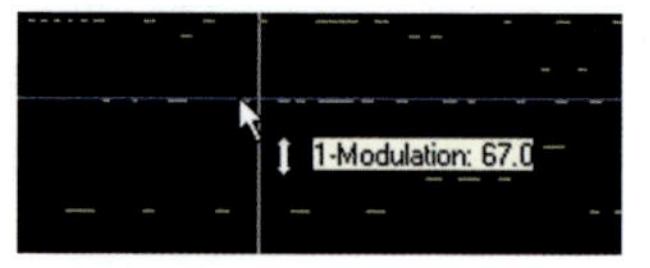

It's possible to record some soft synth control changes depending on the synth itself, if it's possible then it may need to be activated somewhere on the synth so you'll have to check the documentation to find out. If you're not sure then you can just try recording it anyway as follows:

1 Using the MIDI Track that is sending to the synth (usually inserted when the synth is loaded) click the Record Arm button (R).

2 Open the synth by double-clicking its name in the Synth Rack.

3 Click the Record button (●) or press R then tweak your synth!

4 If you had to activate Automation Recording on the synth you should turn it off again when you've finished the recording.

This type of recording shows up as if it was data drawn in the Controllers Pane of the Piano Roll View. If you want to see it as Envelopes then do the following:

1 Click on the Clip containing the controller data to select it.

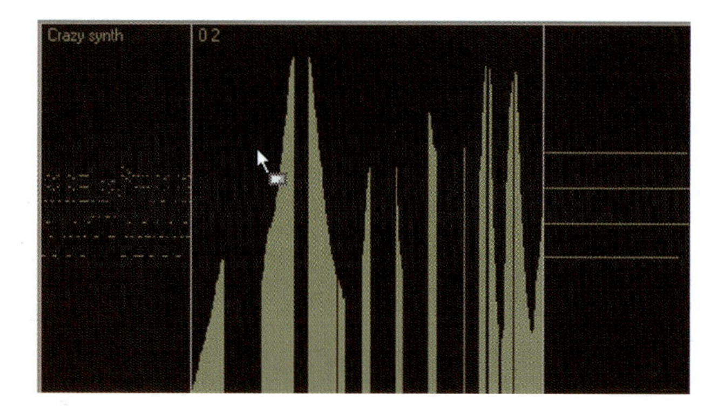

2 From the Edit menu choose Convert MIDI To Shapes (right at the bottom of the list).

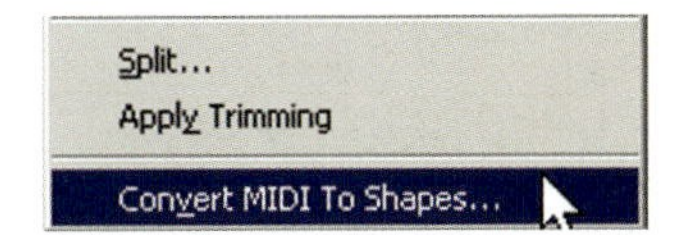

3 Select which controller Type, Value and Channel you want to convert and click OK.

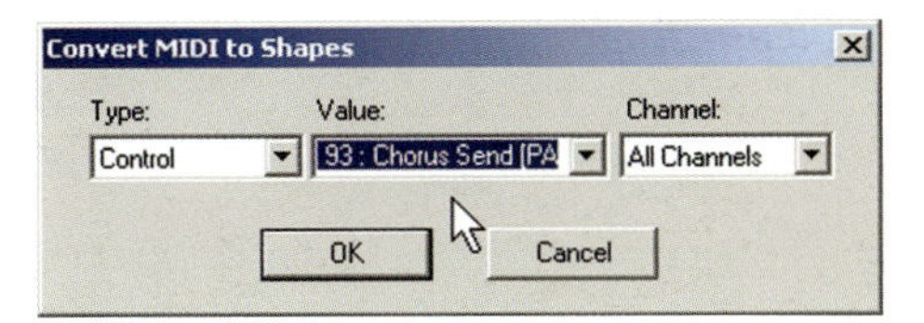

4 The selected data will be converted to an Automation Envelope ready for editing.

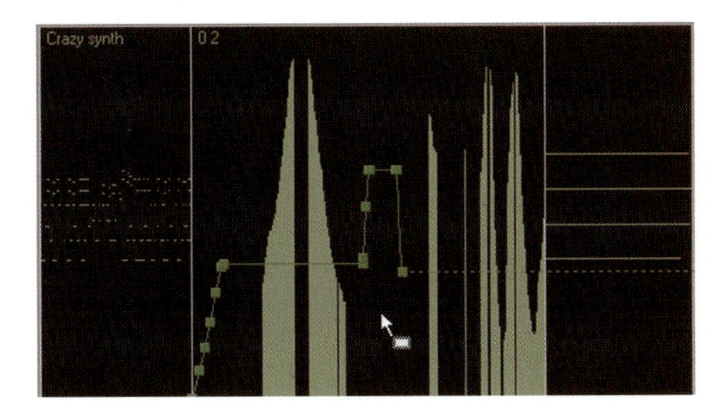

If you wanted to, you could leave the data as it is and open the Piano Roll then edit in the Controllers Pane or use the Event List View Alt+4.

Advanced Envelope Editing

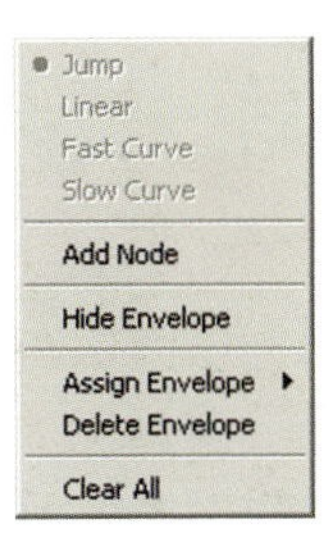

There are a range of useful options available when an Envelope is right-clicked: The options can vary depending on which part of an Envelope is clicked.

A dotted line may denote an area without any data changes, if so some options will be unavailable or 'grayed out'.

An area which has one or more Nodes and a solid line will show all options.

The top section of this menu provides various shapes that can be applied to the Envelope.

Jump
• Linear
Fast Curve
Slow Curve

Jump is most useful for providing accurate switching points for controls such as Mute or Bypass as they will change state when the Envelope crosses the horizontal center line of the Track.

The other options in this section can smooth transitions between Nodes without the need to add more Nodes to create a smooth shape.

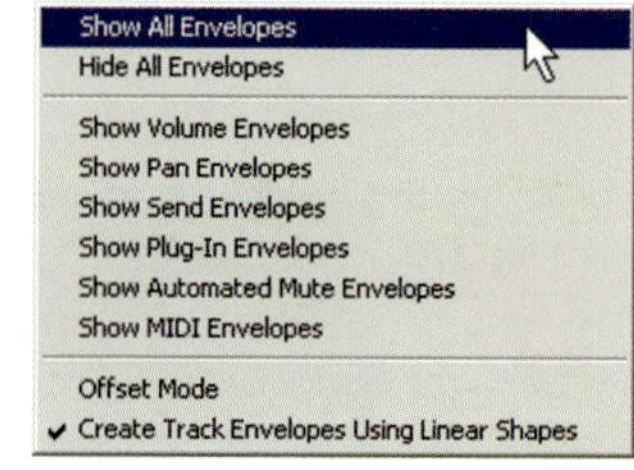

Add Node can be used instead of double-clicking the Envelope and Hide Envelope will obviously hide it! You can show it again by selecting a Show option from the menu attached to the Envelope Tool button.

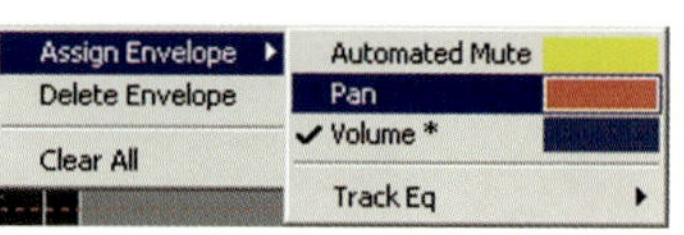

Assign Envelope provides options to assign the Envelope to another available parameter.

Delete Envelope is self-explanatory and Clear All will remove all Nodes from the Envelope and return it to the control's current state.

Right-clicking on a Node also provides options to Reset the Node to its default state or delete it. When selected, Nodes will turn white.

With the Envelope Tool selected you can also drag a lasso around several Nodes and drag them all together as a group.

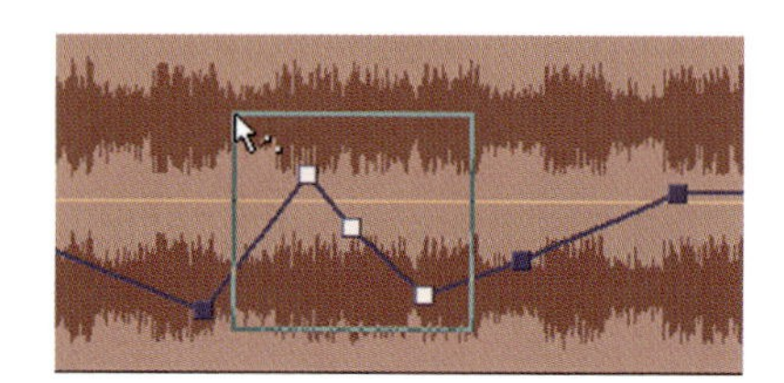

Offset Mode

Available from the Automation Toolbar ( ) the Envelope Options menu attached to the Envelope Tool button or by pressing O, Offset Mode provides another layer of editing that can be performed without disturbing a mix. When activated, controls such as Volume, Pan and Sends will add an Offset to the original control's value. They can be identified in this mode as they gain a + at the left of their sliders in the Track View as an identifying symbol ( ).

The Offset applied is in addition to the current values of controls and doesn't necessarily override it. A Pan set fully left in the normal Envelope Mode would sound as fully to the right if the Offset Pan was set to full right but if the Offset Pan was set at 50% right then the normal Pan would sound in the center as the Offset applied would be 50% of the way from the left, thus central. It may take a little while to figure out but the best thing to do is just use your ears!

It's easy to forget changes made here and it can become frustrating trying to find out why a control doesn't sound how it should if you forget that you've applied an Offset to it so make it a first port of call when things don't sound as you expect them to, you might have tried fine-tuning in Offset Mode a week ago and forgotten about it!

Offset Mode can be useful for getting the rough mix right first then fine-tuning it with the use of Envelopes.

CHAPTER 11

WORKING WITH VIDEO

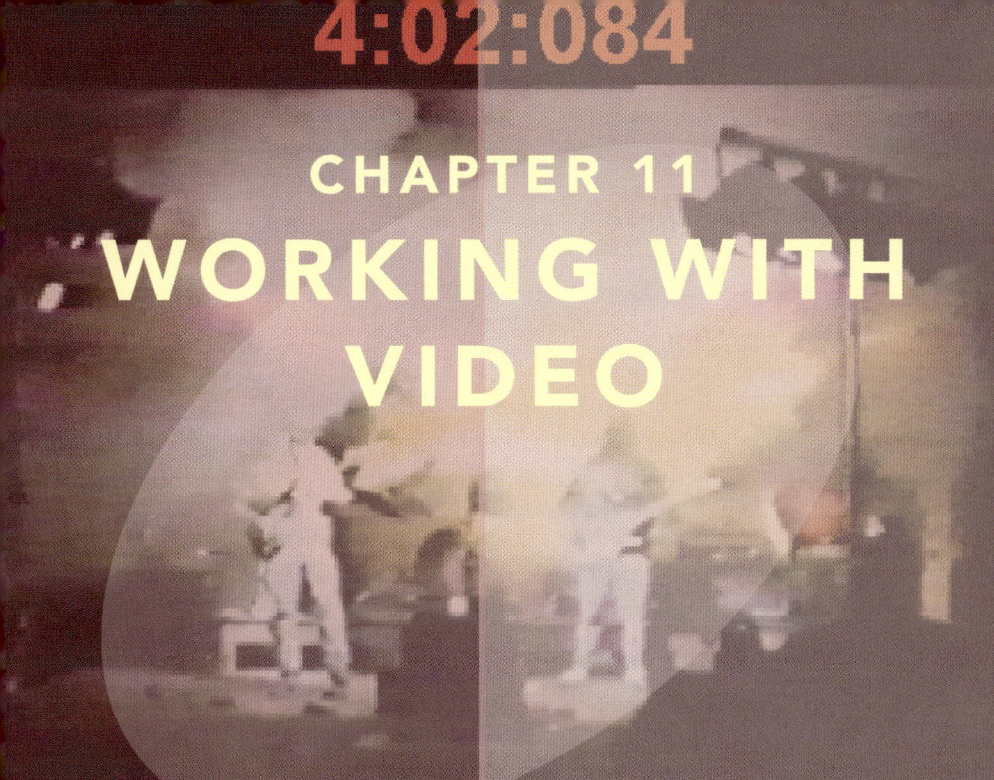

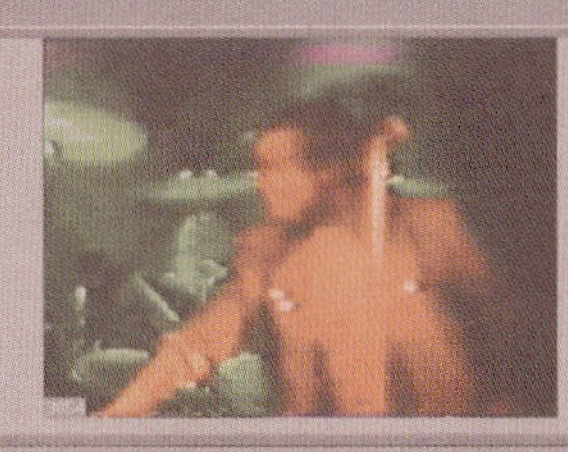

Although SONAR is not a video editing package it does have facilities to enable accurate scoring to video. The resulting material can be either Exported in audio format for use in a dedicated video package or Exported as a video file containing both video and audio. If you Import video files with audio content you can elect to Import the audio with the video and it will then be stripped out and placed into an audio Track within SONAR. This is extremely useful as you can edit and mix this existing audio along with any new Tracks that you may add whatever it contains. If the video contains any sections that you need to sync new sounds with, then the original sound may show some accurate peaks that can be used as a guide for the placement of new sounds.

For proper video use it's a good idea to have your video stored on the system somewhere, preferably on its own hard drive. When backing up video Projects as Bundle files, be sure to save the video separately as it is not saved with the Bundle.

To Open a Video . . .

1 Start with a New or existing Project and from the File menu select Import>Video.

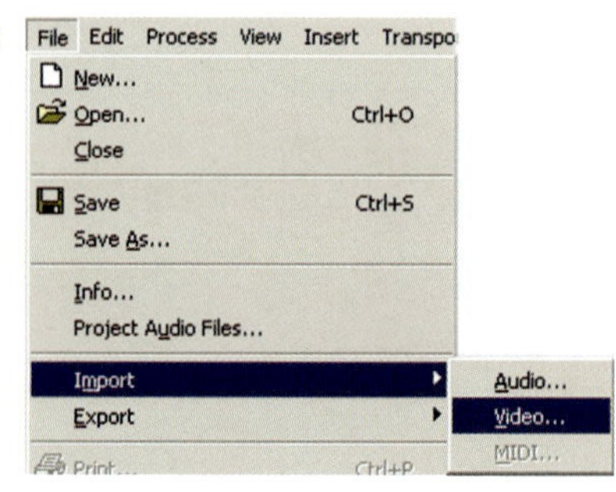

2 A Browser window will appear where you can find your video files, it will open at the location set in Options>Global> Folders under the Video Files path. If your files are usually stored in another location then you can set a new default here to make future searches faster.

3 Depending on the format you are looking for, you can change the Files of Type field in the browser to a variety of formats or if you're unsure then you can choose All Files.

4 If Show File Info is ticked you'll see the file's properties displayed in the File Info box when selected.

File info
Format tag: PCM
Attributes: 48000 Hz, 16 Bits, Stereo
File length: 10356005 Samples (215.75

5 If you want to Import the audio then make sure Import audio stream is ticked; and if you want it splitting into separate mono Tracks, tick Import as mono Tracks.

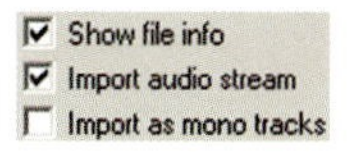

There are two ways of viewing video, the first using the separate Video View available from the View Toolbar ( ), View menu or by pressing Alt+6.

The Video View is a resizable window which has a useful right-click menu that allows a range of options to be set including stretching the video to fit the window (you can also preserve the aspect ratio), altering the time display and background color and viewing extensive video properties. It's also possible to Insert and Delete video from here.

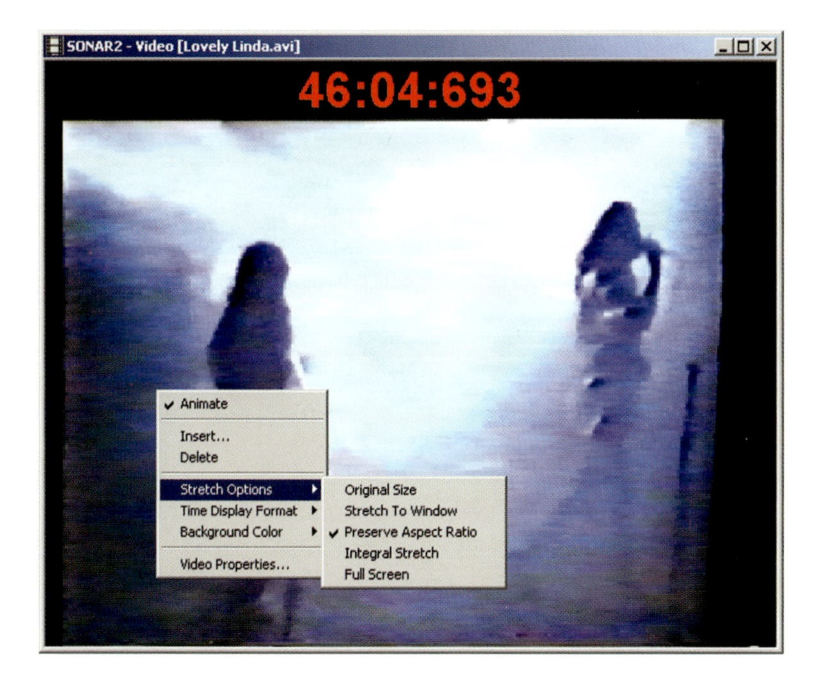

If you do a lot of scoring to video and have two monitor screens on your PC, you could drag this window to your second monitor.

The second method is the video Thumbnails Track along the top of the Track View opened by clicking its button ( ) at the top of the Tracks Pane or by pressing L.

The video Thumbnails Track and Thumbnails Pane is similar to an audio or MIDI Track and will update when you zoom in or out, it can also be reduced to a smaller size using the horizontal splitter bar.

It has two buttons, one to show or hide the frame numbers ( ) and the other to show or hide the actual video frames ( ). You can also change the video start time by applying an offset in the Start field if necessary (Start 1:01:000).

If you right-click on the video Thumbnail Clips Pane you can choose to see Absolute Frames, view the Video Properties or go to the Export Video dialog.

When working with video a useful tool is the Scrub tool, available by clicking its button ( ) or by pressing B. The tool can be used to drag along the Tracks to hear the audio and see the video in the Video View update in real-time. This is great for accurate syncing of sound to picture.

Dragging on a Track will play just that Track, while dragging in the Time Ruler will play all Tracks simultaneously.

If you click in the Time Ruler the Now Time will jump to that position (assuming Snap To is off) but if you click on a video Thumbnail it will jump to the start of the displayed frame.

Your music and sound can be Exported using the normal methods but you can also Export the complete video from the Export dialog in the File menu.

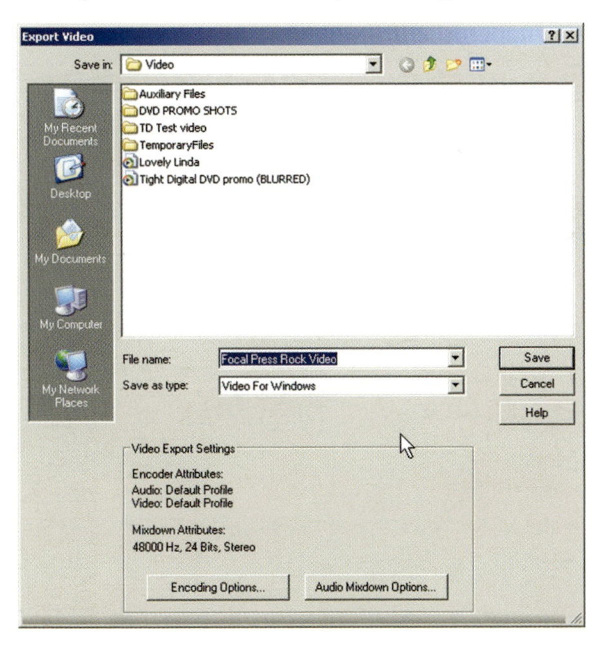

Using this method offers a variety of choices including the type of video file to Export as, its Encoding Options and also the Audio Mixdown Options. You must make sure that you have selected all of the audio that is to be included with the file before you Export it.

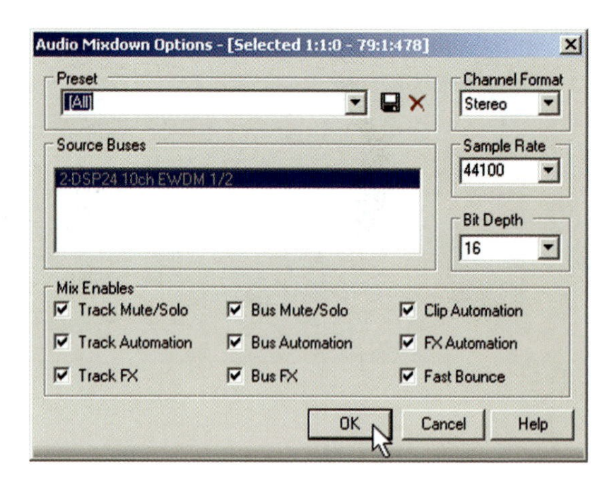

CHAPTER 12

MIXING DOWN AND MASTERING

It's not an easy task to explain how you should mix or master your music as there are so many variables and you should have a good idea how you want the music to sound anyway. This chapter isn't about what you should do but more about how you can carry out many of the processes involved in mixing and mastering your music. It's easy to make mistakes and a little help can go a long way so here are some guidelines that should help you to produce good quality music without too much of a struggle but as with most things in life you will improve with practice as there's no substitute for experience.

Inboard or Outboard Mixing?

If you're used to using studio hardware then your familiarity or preference may well mean that you prefer to mix outside of the computer which is fine. As you get used to the program you can introduce various aspects that make life easier, such as Automation, a little at a time.

I personally use a bit of both as my computer never seems to have quite enough power to handle everything I want to do (must get a new one) so I tend to send Subgroups and important Tracks out to my hardware in order to use mixing and processing tools without hitting my processor.

I usually Subgroup instruments such as drums, guitars and vocals then I can send the Subgroup outputs directly to my mixer where they can be treated and controlled externally. I use a digital mixer so I can also save all of these settings and back them up along with the SONAR files. I do use a lot of processing in SONAR but it depends on the Project, particularly the number of Tracks and their requirements. As a general rule I will EQ, compress and effect Tracks as necessary in SONAR and process their Subgroups externally but this is flexible and I might take one or two individual Tracks outside as well. Small Projects will invariably stay in SONAR for everything.

Here the lead vocal channels on the left are all routed into the Vocals Subgroup on the right which goes out to its own stereo audio channel routed to my hardware mixer and outboard effects.

Basic Guidelines

Mixing involves making each element of your music sound its very best and fitting it into place with all of the other elements which is not always as straightforward as it sounds. You should be thinking about this as you actually record your songs and make suitable provisions as you go such as trying to use instruments that will complement each other and not fight for space in the mix. Several sounds in the same frequency band will be very difficult to define so if you have a bass drum, bass guitar and bass synth line then do try and make them sound different to each other right from the outset, it really will make life easier later on in the Project.

If you record audio at high levels then you'll have cut down on any background noise which will be of benefit when mixing down, especially if you add compression to the audio which will increase any noise present.

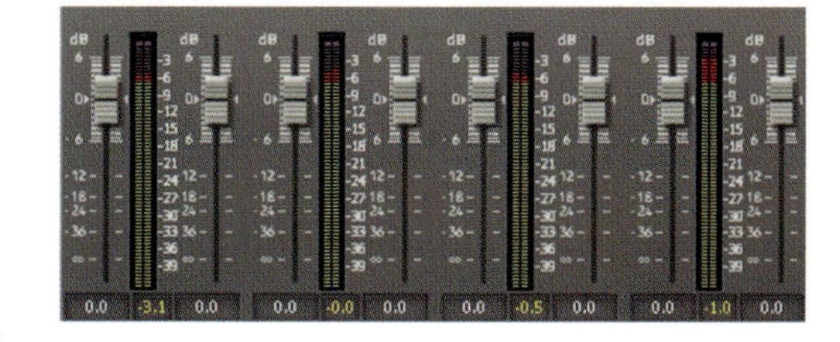

Volume and Trim controls should be kept in order and also an eye kept on the Main Outs as the cumulative signal level is likely to peak and clip when several individual channels are increased. Consider reducing levels and possibly introducing compression or limiting if necessary to control this.

If your CPU meter is running hot then freeze any soft synths that you can and if necessary Bounce any Subgroups of Tracks to a new Track then Archive the originals and Hide them.

Think sensibly about creating space for sounds by using the full stereo width and using Pan settings intelligently ( ). Panning sounds properly can often clear up a cluttered mix without having to perform any serious tweaking on the sound itself.

Use EQ to 'clean up' sounds. If a piece of audio shouldn't have any bass frequencies present then you can roll them off to ensure that any spurious data at these frequencies is eliminated and can't cloud the mix. Low Shelving or Highpass filters can be used to cut out these frequencies and save speakers from working harder than necessary reproducing them.

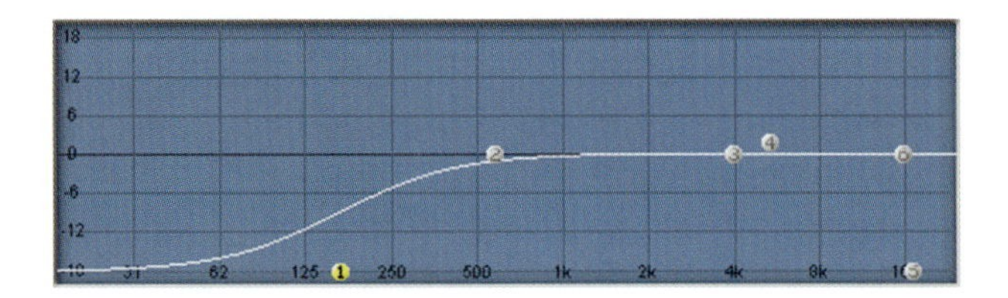

Try to reduce not increase settings to tidy the mix up; is the piano really too quiet or are the guitars too loud?

Aim to produce a mix that shouldn't need anything doing to it at the mastering stage.

Leave some time between recording, mixing and mastering if possible, the longer the better really as you'll be able to judge the sound more critically with fresh listening.

Mastering is making the most of your mix so if your mix is really good then you won't have to spend too much time mastering it!

Use a commercial CD in a similar style to your own song as reference when mixing and mastering. Just listen to it occasionally to see if you're achieving the same level of quality and sound.

Try your mixes out on as many different setups as you can such as in the car, on a personal stereo, on a hi-fi and on a large PA system if possible.

Markers

Not exactly a mixing tool but very useful nonetheless, Markers can be placed anywhere in a Project either on-the-fly or when the song is stopped and can be placed at any stage of the Project. They can be named for fast identification and moved around to suit edits made within the Project. They also allow rapid location of specific points when you use the Markers Toolbar to jump between them (Key change C).

To insert a Marker when the song is stopped:

1. Place the Song Position Pointer where you want the Marker by clicking at the appropriate point.
2. Click the ( ) Insert Marker button in the Markers Toolbar or press the F11 key.
3. This dialog will appear where you can type in a name for the Marker (such as Verse 1, Solo, etc.).

4. If necessary you can confirm the Time setting then hit OK.
5. A Marker will be inserted at your chosen point and named accordingly.
6. If you don't type a name, the Marker will still be inserted without a name.

To insert a Marker when the song is playing or recording:

1. When the Song Position Pointer reaches the appropriate place hit the F11 button.

2 A Marker will be inserted and named with a letter and number (e.g. A1) ( ).

3 You can continue to place Markers at any point in the song using this method as they will be named consecutively (A2, A3, etc.).

To name the Markers after stopping playback or recording:

1 Make sure you have the Markers Toolbar open. View>Toolbars then tick its box.

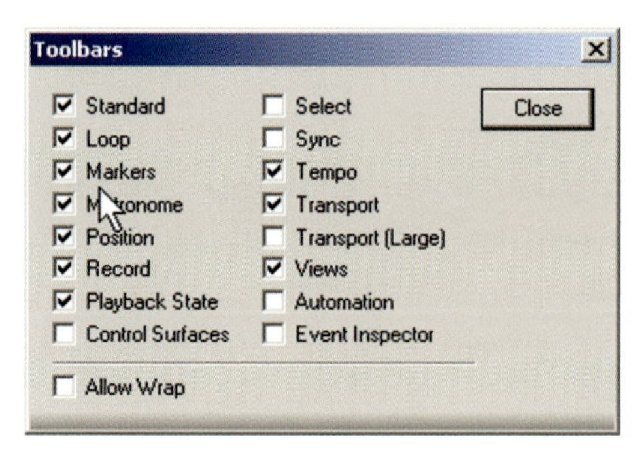

2 Click the ( ) button to open the Markers View. (This button is also available from the Views Toolbar.)

3 The Markers will be shown in a list, double-click on any one and the Marker dialog will open where you can type a new name and adjust the Time if necessary.

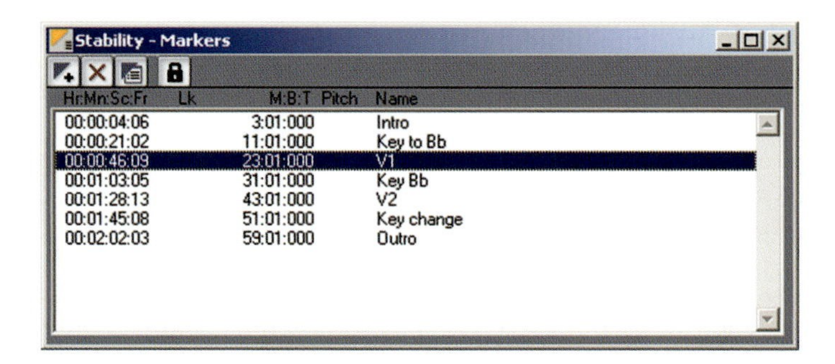

4 Alternatively you can right-click on a Marker flag to invoke the Marker dialog.

To jump between Markers use the ( ) button to jump backward one Marker and the ( ) button to jump forward one Marker. You can also select a Marker's name from the drop-down list to jump directly to it.

Pressing F5 twice invokes a dialog which let you select any Marker from a list and jump to it.

Clicking between two Markers in the Time Ruler selects the region between them.

Good Habits

When laying down your recordings there are things that you can do which will help when you come to the mixing stage but if you missed any out then now's the time to do them. Some people will mix as they go but others prefer to record all of the necessary parts before tweaking, there isn't a right or wrong way, just use whatever suits you best. Here are some helpful hints to make mixing a bit easier.

Firstly it's very useful to name each take and also to number any multiple takes such as Guitar Solo 1, Guitar Solo 2, etc.

I also like to group similar Tracks together so all of the individual drum parts will be in a logical group. This is easily achieved by dragging them up or down. You can also use Track Folders to organize them efficiently.

Archive any Tracks that definitely aren't going to be used right away ().

Hide any unnecessary Tracks by using the Track Manager; press M to open it; alternatively select the Tracks to hide and press Shift+H to hide them. Pressing A will show All Tracks.

Create Subgroups when possible.

Tidy up any starts, endings and sections where musicians are 'laying out' with Slip Edits and Fades.

If your soft synth parts are finished then use the Freeze function to convert them to audio and free up your resources.

Decide what effects you are going to use and if any of them will be shared among several Tracks, if so then they should be on an Aux Bus.

Use Automation Envelopes to perform minor changes such as a dip in Volume.

It's always worth making a backup of your Project at this point, particularly Save As a Cakewalk Bundle file that can be recovered should you decide to go back to the basics. Choose Save As from the File menu and select Cakewalk Bundle in the Save as type field. Give the file a suitable name such as 'Pre Mix' and if you make more than one backup give each one a logical number such as 01, 02, etc. These files can be written to CD or DVD discs for long-term storage.

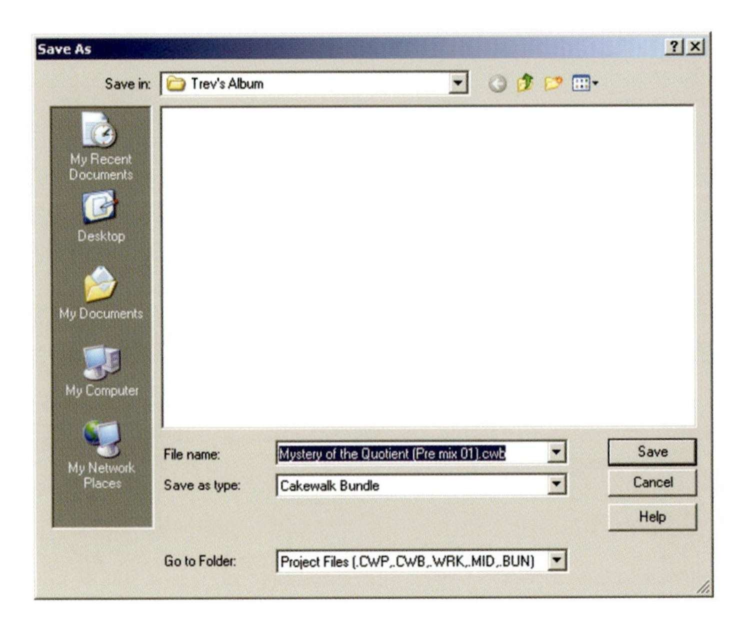

EQ Tweak

The EQ is inserted before the effects by default but at times you may want to place it after (Post) effects maybe to control a change caused by a plug-in. To do this just right-click on the EQ Plot ( ) and select Eq Post Fx to change it.

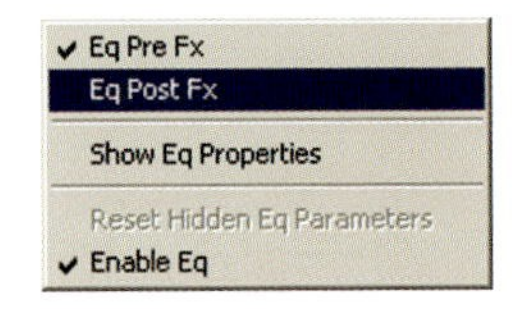

Track Folders

When you have a lot of Tracks to mix you may need more room onscreen so it's a good idea to put logical groups together. This can be extended by the use of Track Folders which enable you to store multiple Tracks in the same space as a single Track but a single click opens up the Track Folder to provide instant access to them all; you can also control some parameters of the whole group in one go by selecting the Folder then performing the change. Here's how to use Track Folders . . .

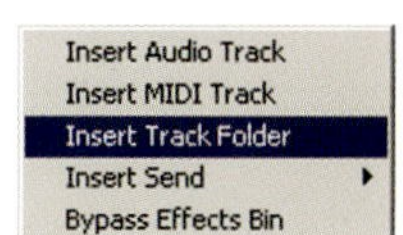

1 Right-click in the Tracks Pane (on a Track number will do) and select Insert Track Folder.

2 Double-click on the Folder name and type in a suitable new name.

3 Select which Tracks will go into the Folder Track by holding Ctrl and clicking on each one's number.

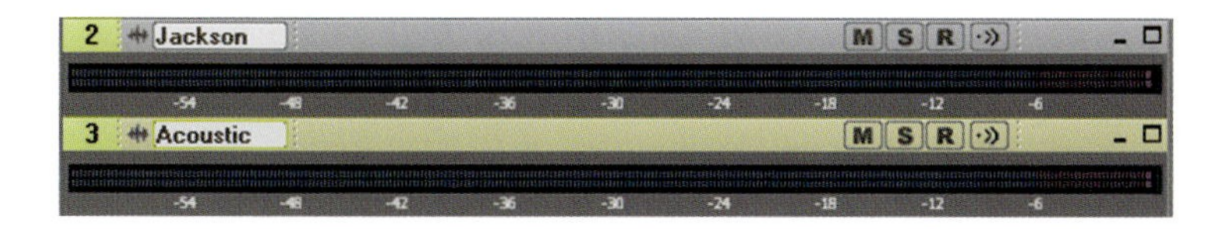

4 When you've selected them all right-click on one of the highlighted numbers and choose Move to Folder then select the name you just gave the Folder. Alternatively you can highlight the Tracks then right-click on the folder and select Add Tracks to Folder. You can add other Tracks later using either method.

Move To Folder
Remove From Folder
Guitars
New Track Folder

5 The Tracks will now be part of the Folder Track and you can see the line linking them. You'll also see that the Folder shows the number and type of Tracks that it contains.

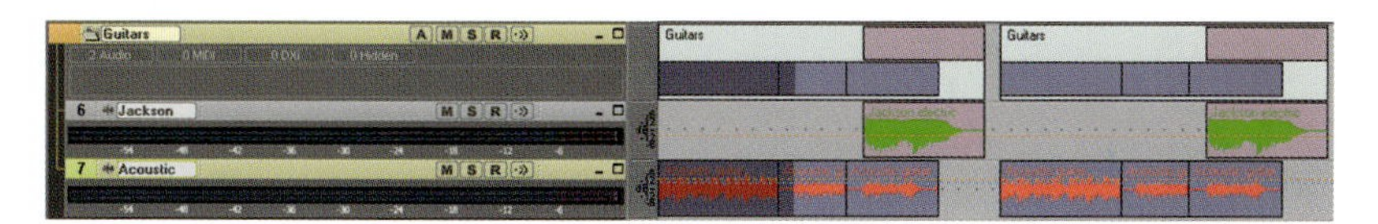

6 Each Track can still be resized, edited, effected and controlled in the usual manner but you also have control over the whole Folder. Clicking on the Folder icon will open or collapse the Folder allowing you to view the Tracks it contains or hide them all in one go.

7 The Folder Track also has a set of controls to Archive, Mute, Solo and Record Arm all of its Tracks with a single click (A M S R).

8 You can also perform Slip Edits and apply Fades to all Tracks within the Folder by selecting the Folder Track (either click on its 'Composite Clip' or the space to the left of the Folder icon) then performing the edit on any of the highlighted Tracks within its group (you can perform Slip Edits on the Composite Clip itself but not Fades).

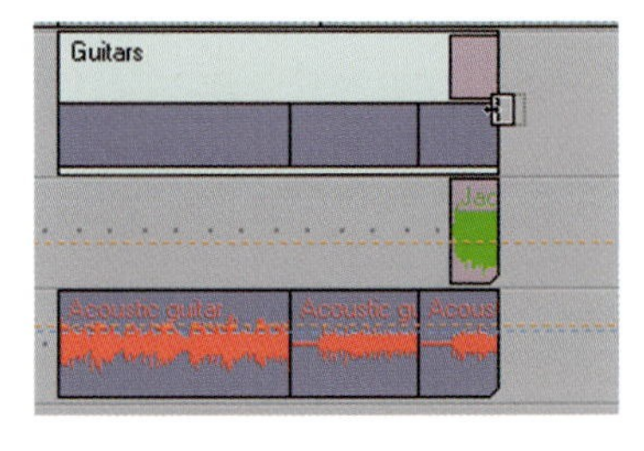

9 Tracks or Clips can be removed from the Folder by simply highlighting them and dragging them out.

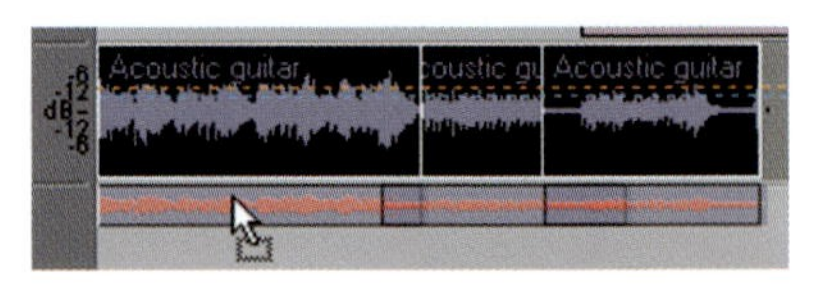

10 The Folder may be deleted with or without deleting the Tracks it contains by right-clicking to the left of the Folder icon and choosing Delete Track Folder.

11 A dialog will appear asking if you want to delete the Tracks within the Folder as well, just choose No to keep them.

Comping

When you've recorded numerous takes of a piece (a lead vocal for instance) you'll need to compile the takes into a Track that includes the best parts from each performance; SONAR has a selective Muting facility that makes this job a little easier.

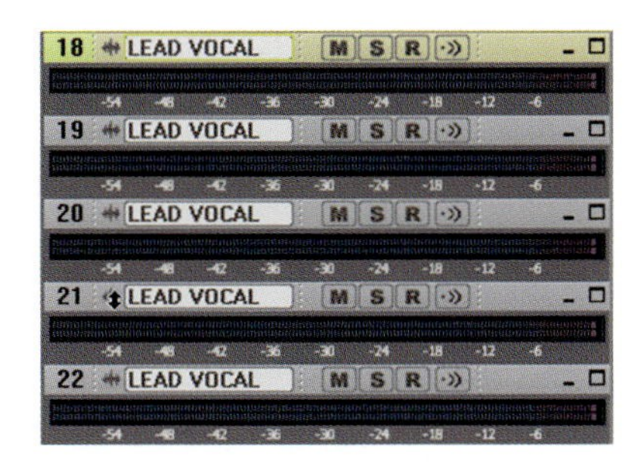

You should organize the takes so that they are all adjacent to each other first, either in adjacent Tracks, as Layers in a single Track or in a Track Folder.

Expand them so that they are easy to see and work on then . . .

1 Select the Mute tool from the toolbar at the top of the Tracks Pane by clicking it or press K (⊘).

2 The menu attached to the button can be used to make the tool Mute Time Ranges or Entire Clips. We're using the former so make sure that it is selected.

3 Solo each part and listen to see which is the best take of it.

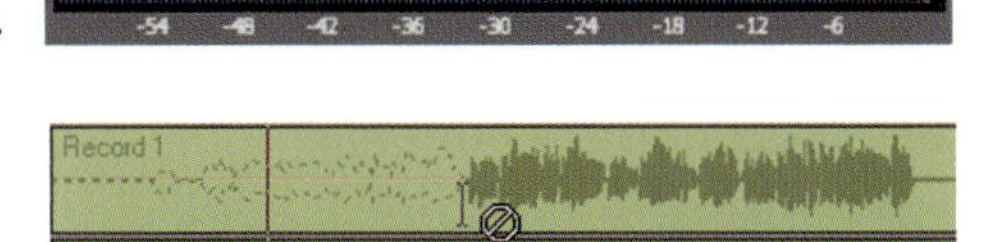

4 When you've decided which take of a particular part sounds best you can selectively Mute the other Clips in just that area by clicking and dragging in the lower half of each Clip. The part you dragged over will be Muted.

5 If you make a mistake or change your mind then just click and drag in the upper half over the area you want to Unmute and it will be restored.

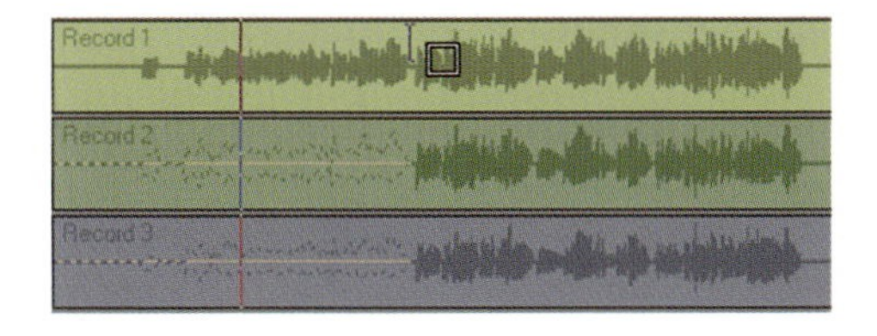

6 If things are a little tricky then use the Zoom tools to get close in.

If you have Clips in Layers on a single Track then holding down Ctrl when applying Muting or Unmuting with the Mute Tool will Mute the region in all overlapping Clips in the same Track.

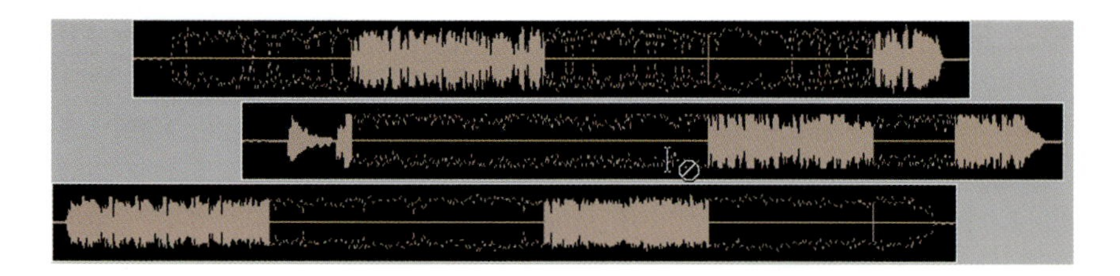

Using Another Audio Editor

Although you can perform pretty well anything you like on your audio files within SONAR itself you may own a wave editing package such as Sound Forge that you like to use for certain procedures. If so then you may occasionally want to use it to work on an audio file in a SONAR Project. You will find that if the program contains compatible plug-ins they will already be available in SONAR and can usually be operated without problem. If however you want to use the wave editing program itself try this . . .

1 In SONAR click on the audio Clip to be used so that it is highlighted.

2 Go to the Tools menu and you should see your wave editor's name, click on it.

3 The wave editor will open and the audio will be automatically loaded in ready for editing.

4 When you've finished, Save then close the editor.

5 SONAR will now ask you if you want to reload the changed audio, click Yes (or Yes all for multiple selections).

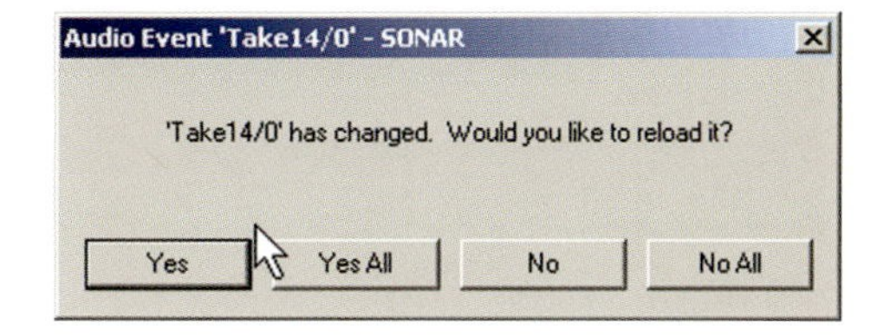

6 The audio is placed back in the correct position with the edits applied.

If your wave editor is not available from the Tools menu then you can always Export the audio in question and then Import it back in. If you do this it's a good idea to Bounce it down so that it starts at the very beginning of your Project so that it is easy to line up when you Import it again or if your wave editor supports them you can Export in the Broadcast Wave File format.

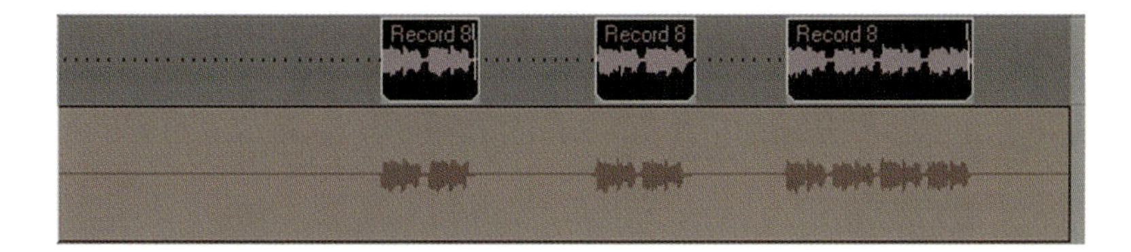

Bouncedowns

If you've ever worked with small multi-track tape machines you'll probably be familiar with bouncing down as it used to be the procedure necessary for freeing up tracks by 'bouncing' several tracks onto a single or stereo pair of tracks thus freeing up their original space for more recording. The big drawback with this was the loss of quality resulting from the Bouncedown which got steadily worse during any subsequent bounces. This is no longer a problem as any bounces performed within the digital domain are exact copies of the originals and it's also easy to keep the originals just in case you need them later.

There are many reasons that you may want to perform a Bouncedown such as a submix of Tracks (maybe the rhythm section for a soloist to rehearse to), to render a synth as audio or just to relieve the pressure on your system when using a high number of Tracks. The principles are basically the same and you can (and should) always keep the originals.

To Bounce down a complete mix use exactly the same principle but you would obviously include all of the Tracks in the song not just a selection of them. Any Tracks using soft synths or ReWire devices will be automatically rendered as audio during the Bounce. The only thing you need to remember is that if you are using any external processing then this won't be included so you must record those Tracks as audio before you perform the Bounce.

To Bounce several audio Tracks down to a single stereo Track . . .

1 The outcome will depend on what you select in the Bounce dialog box so the easiest way to do this is to get the Tracks sounding how you want them to and Mute any that you don't want to include. Here we're going to Bounce the drum Tracks so all the other Tracks are Muted.

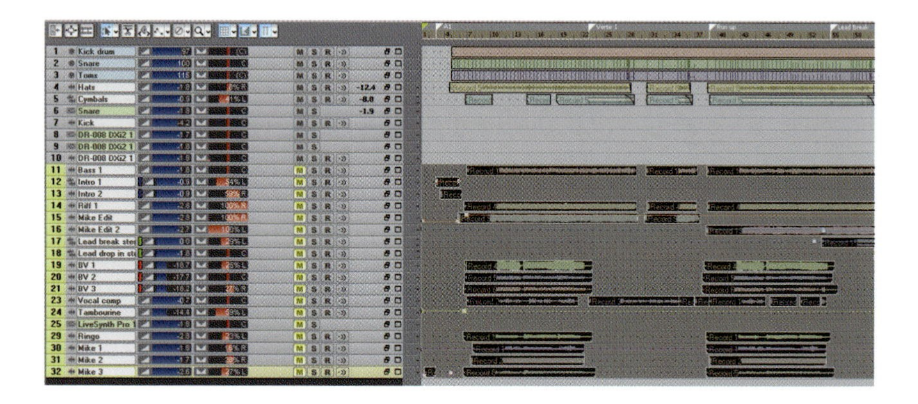

2 Select all of the Tracks then go to the Edit menu, select Bounce to Tracks and choose a Destination for the new Track or Tracks. You will be offered a new Track which is usually the best option.

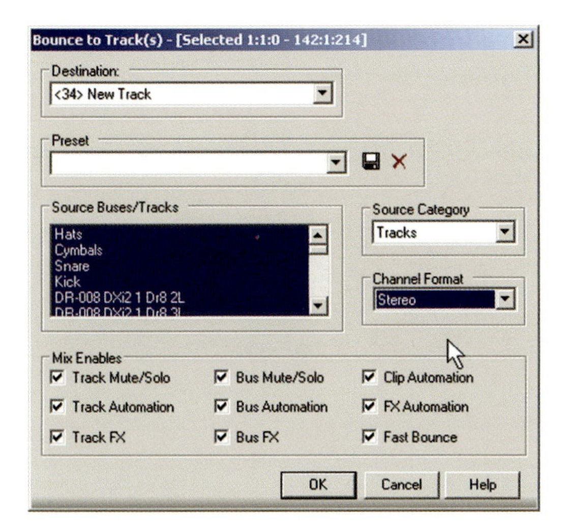

3 Choose a Preset, in this case we're using the [All] option. The other fields will change to reflect your choice here or you can change them to suit your needs, note that the Source Category in conjunction with the Source Buses/Tracks fields will produce separate Tracks for each selected Source so if for instance you set a Tracks Category and highlight several Tracks in the Source Buses/Tracks field they will be Bounced down as separate new Tracks not as a single new Track.

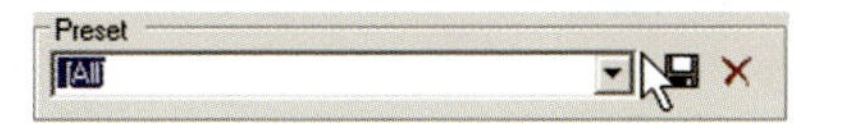

4 Choose your Mix Enables settings, in this case we're leaving them all checked as we want to hear everything exactly as it is including any settings and Automation.

5 Click OK and the audio will be rendered as a new stereo Track and named accordingly.

6 Now you can select all of the Tracks that you used to perform the Bounce and Archive then Hide them. Don't forget to unmute the others.

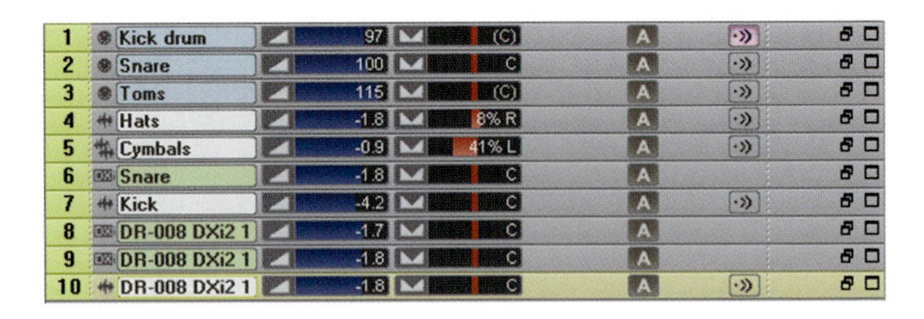

This method can be used to create mixes of any Tracks within your Projects which can be very useful for making several versions quickly. This can be particularly useful for providing musicians with a mix that doesn't include their instrument so that they can practice or even record their own part separately, in another studio maybe. The new part can then be Imported into SONAR at a later date and mixed in to the existing song. Using this technique enables you to send mixes to anyone in the world as you can email the audio or Export as a compressed file type such as mp3 which will make the file size more manageable. Note that you may have to upgrade or purchase an encoder for some file types, a dialog will inform you if this is the case.

Checking the Final Track

Here's a way to check if your final mix is too loud and causing clipping to occur and how to get round it before Exporting it. You need to route all Tracks and Buses through a single Main or Bus for it to work . . .

1 Open up the Console View and drag open the Main or Bus area so that you can see the meters of whichever Bus the Tracks are running through.

2 If necessary lock the pair of faders together with the Link Faders button ( ).

3 Play the song from start to finish.

4 Now look at the Meter Peak reading figure below the relevant meter (3.5).

5 If the reading is red then your song has gone over so you should pull those faders down by the same amount; for example if your Peak is 3.5 then pull them down until they show −3.5 or a little less.

6 Your song will now play without clipping and can be Bounced or Exported safely.

7 If you perform a Bouncedown first then don't forget to reset the faders back to 0dB after or you'll be reducing the signal twice!

Note that you can only really use this trick if the level is due to the accumulation of signals. If you have a Track or two clipping anyway you should consider reducing them first.

Exporting Your Work

When all is complete you'll want to save your efforts onto a CD or share them with others as mp3 files over the Internet or even use them as a film score or DVD surround soundtrack. Whatever the reason you'll have to Export the finished audio.

You can Export without Bouncing down to a stereo Track first if you prefer but I always like to make sure that everything sounds just right first before Exporting on finished material, I do sometimes Export without Bouncing when I'm just trying a mix out, then I can quickly burn it onto CD and listen to it on a few systems before making a final decision.

Exporting is very similar to Bouncing as you need to make sure you have the correct settings to ensure that the Export contains what you want. It works like this . . .

1 From the File menu choose Export>Audio to open the dialog box.

2 Choose the location you want the audio to go using the Look in field and its tools (Look in: New songs).

3 You can now choose a Preset or you may prefer to set the parameters yourself and select which type of file you want it Exported as.

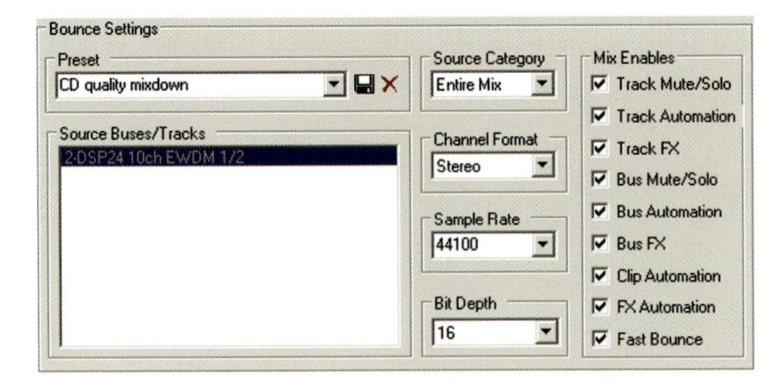

4 If you choose to create a custom setting then you can type a name for it in the preset field and click the Save icon to save it for future use.

5 The Bounce Settings work in the same way as performing a Bounce; to keep it simple the What You Hear Preset is an obvious one to use.

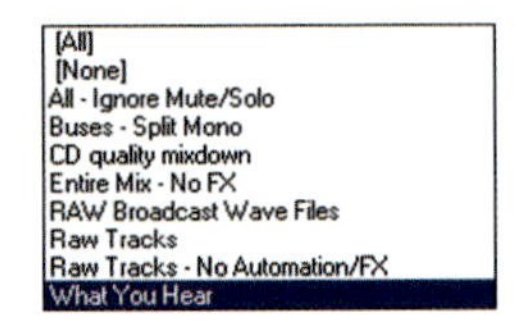

If you wanted to Export all of your Tracks separately then you'd need to select them all first then open the Export dialog and choose Tracks as the Source Category.

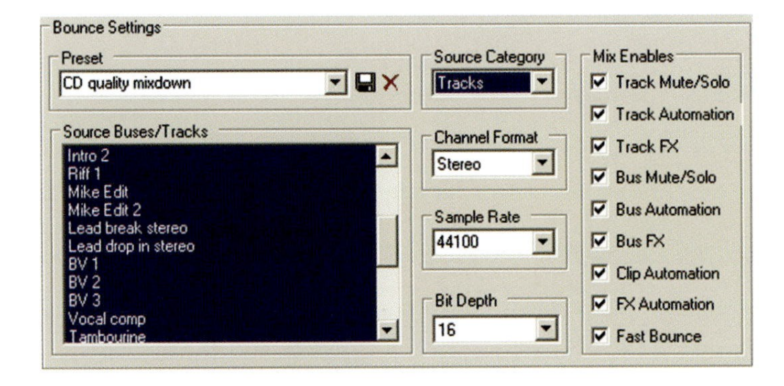

The Source Buses/Tracks field will now show all of your Tracks; type a name in the File Name field and you can then Export them. A dialog will inform you what is going to be Exported and ask you to confirm this.

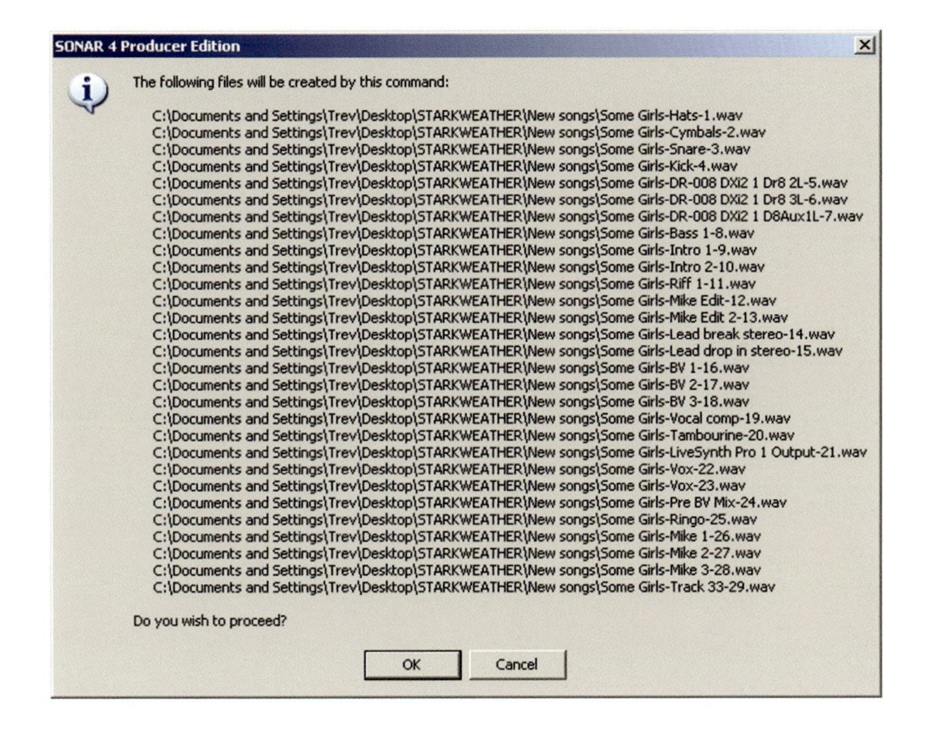

CHAPTER 13

SONAR PRODUCER EDITION VERSUS STUDIO EDITION

SONAR is available in two editions, the Producer Edition which is the top of the range, flagship product and the Studio Edition which is a cut-down version of Producer. Although the Studio version is an excellent package it does have some important differences that are outlined in this chapter. You may not need or want these features in which case Studio will do a fine job of most duties anyway but if you find that you do need those extras then you can always upgrade. Check out the latest information at www.cakewalk.com for details.

The surround sound features are not available in the Studio version which may not be a great loss to you as it is still quite a specialized area and requires all the extra hardware such as speakers and amplifiers as well as an encoding system to be properly effective. It is however a technology for the future so if you like to keep up then it is worth considering.

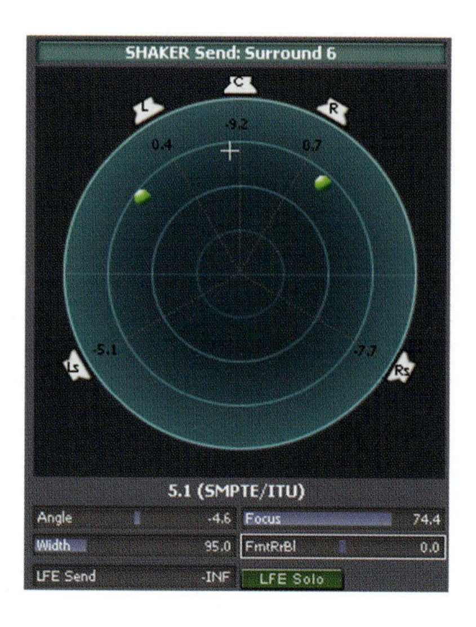

Along with the surround feature you also lose the surround effects that are included with Producer Edition as they wouldn't be much use without the other facilities anyway! You do get a cut-down version of the Lexicon Pantheon Reverb (Pantheon LE) but Producer owners get this full-blown surround version.

The high-end dithering and time-stretching technologies aren't built into the Studio Edition which means that if you work at high resolutions then Export at a lower resolution (for CD maybe) then you could be losing some quality. If you Export at a high resolution then convert with a quality external tool such as those built into wave editors then this might not be a problem and if you don't work at high resolutions it probably won't matter anyway.

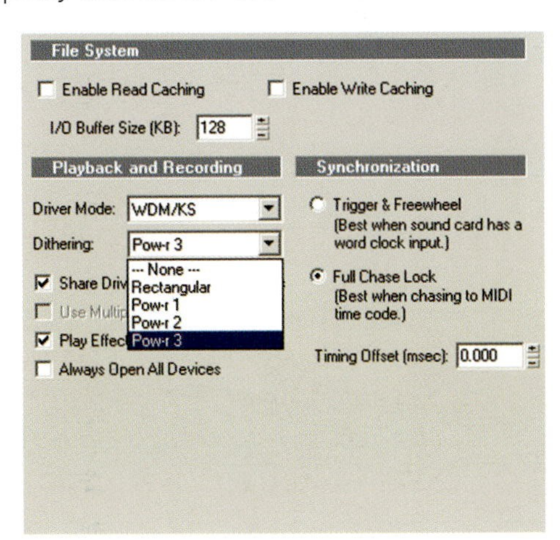

The loss of the Sonitus effects suite from Studio Edition is a bit of a blow as these really are quality plug-ins and I'd suggest are well worthy of consideration if you don't already have a top-notch set of effects. Here are some of them. . . .

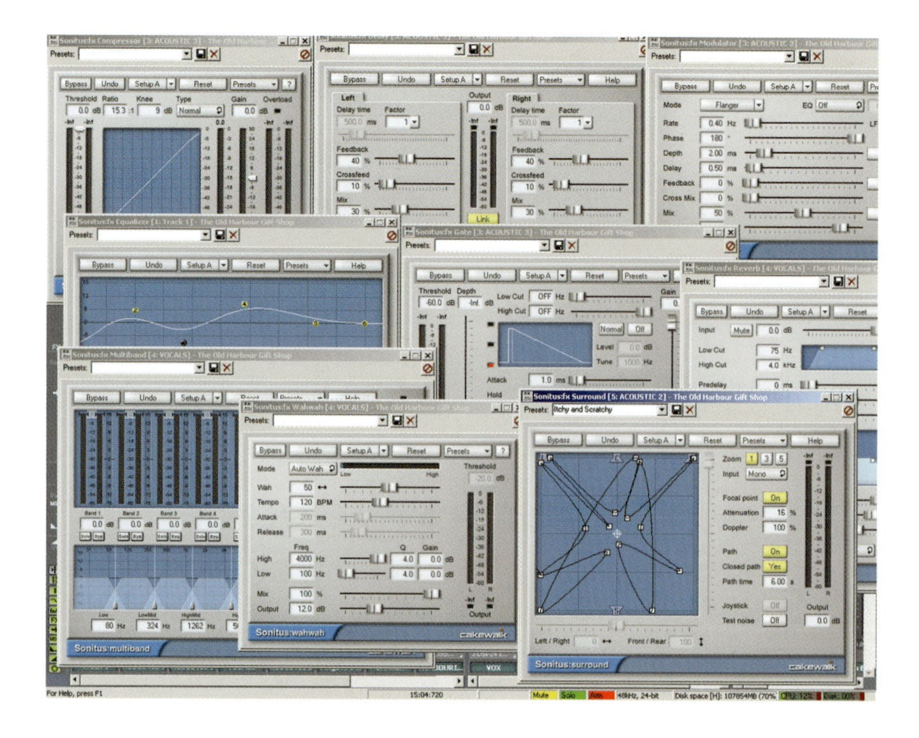

You don't have the built-in EQ in the Studio package which provides a ready inserted EQ, the Sonitus one to be exact, in each audio channel and access to many parameters directly from each channel strip Inspector. This is a very useful feature and ergonomically speaking can save a lot of time and hassle. You also lose the ability to access some effect parameters directly from the channel strip Inspector.

If you use video a lot then the loss of the video thumbnails Track may be something to think about as only the Producer Edition has it.

Index